T0239440

Introduction to Continuum Mechanics for Engineers

Nik Abdullah Nik Mohamed

Introduction to Continuum Mechanics for Engineers

With Solved Problems

 Springer

Nik Abdullah Nik Mohamed
Faculty of Mechanical and Automotive
Engineering Technology
Universiti Malaysia Pahang
Pahang, Malaysia

ISBN 978-981-99-0813-4 ISBN 978-981-99-0811-0 (eBook)
https://doi.org/10.1007/978-981-99-0811-0

This Springer imprint is published by the registered company Springer Nature Singapore Pte Ltd.
The registered company address is: 152 Beach Road, #21-01/04 Gateway East, Singapore 189721,
Singapore

*This book is dedicated to Nadzirah, Faiq,
Kamilla and Khalissa*

Preface

The underlying motivation for writing a book on continuum mechanics is to help post-graduate students to understand a little bit more about the concept of continuum mechanics before they jump into the freezing water of finite element analysis, boundary element method or any other computational method available. Most of these students are very well versed and are familiarized with commercial FEM or any other computational method-related software, but their knowledge or understanding of continuum mechanics, which constitutes the basis for complex material analysis, is limited to the accompanying manuals of highly priced software licenses. It is acceptable for practice engineers or for the people who use this sophisticated software for a living but not necessarily so for the people who are doing research trying to discover new knowledge or a new horizon.

The main chunk of this book was finished during the period of Malaysian MCO starting March 2020. It was drafted based on my lecture series on Continuum Mechanics given to post-graduate students pursuing Ph.D. here in UMP and young UMP lecturers just coming back from overseas after completing their Ph.D. abroad.

This book is organized not to provide a very comprehensive and going-to-the-depth Continuum Mechanics but rather to offer a platform for discussion and debating the concept of continuum mechanics in the application of material behaviour analysis, particularly for engineering students or for engineering-related professionals. It is also intended to provide a steppingstone for further study on much more complicated and complex material behaviour. Creep, damage mechanics, crack propagation, i.e. the behaviour of real material with defects and discontinuities are some examples to be named.

The author likes to pay gratitude to those who supported either actively or indirectly completing this book. Many thanks to those who were able to attend my lecture on Continuum Mechanics at the UMP prior to the lockdown. Their interest and eagerness to know more about continuum mechanics indeed fueled my motivation to metamorphize my inadequately prepared lecture notes to this small booklet.

Rinching, Selangor, Malaysia
2022

Nik Abdullah Nik Mohamed

Contents

About the Author

Nik Abdullah Nik Mohamed is a professor in Mechanical Engineering at the faculty of Mechanical and Automotive Engineering Technology, University of Malaysia Pahang. Nik Abdullah graduated with Bachelor of Engineering degree from the University of Applied Science Munich, Germany, in 1979 and received both his Master degree (Diplom-Ingenieur) and Ph.D. (Dr.-Ing) in 1983 and 1991 from the Technical University of Berlin, where he was appointed as a research and teaching associate by the university from 1985 to 1992. In Malaysia, he began his lecture-ship career at the National University of Malaysia (UKM) in 1992 and continued his academic service in the year 2013 at the University of Malaysia Pahang (UMP). Prof. Nik Abdullah's research focuses on the application of continuum mechanics in describing the mechanical and thermodynamic properties of materials and in partic-ular of smart materials, auxetic materials and bulk metallic glasses. In addition, he also actively involved in research and teaching in the area of mechanical vibration and reliability engineering. He has published over 100 research papers, three chapters and two research books.

List of Symbols

Chapter 2

F	Deformation gradient
H	Displacement gradient
C	Right Cauchy-Green tensor
B	Left Cauchy-Green tensor
E	Lagrangian strain tensor
$\hat{E}$	Linearized Lagrangian strain tensor
e	Eulerian strain tensor
I_A	First invariant of the tensor $\mathbf{A}$
II_A	Second invariant of the tensor $\mathbf{A}$
III_A	Third invariant of the tensor $\mathbf{A}$
L	Velocity gradient tensor
D	Rate of deformation tensor, Stretching tensor
W	Spin tensor, vorticity tensor
$D_J\Phi/Dt$	Jaumann time derivative of Φ
$D_O\Phi/Dt$	Oldroyd time derivative of Φ

Chapter 3

b	body force per unit mass
t	traction or surface force per unit mass
S	Cauchy stress tensor
T	First Piola-Kirchoff stress tensor
$\tilde{T}$	Second Piola-Kirchoff stress tensor
Ω	Damage tensor
Ψ	Continuity tensor

Chapter 4

ψ Strain energy density function
ψ Helmholtz free energy function

Chapter 5

b   body force per unit mass
t   traction or stress vector per unit mass
ρ density

Chapter 1
Introduction

1.1 Description of Material Behaviour via Continuum Mechanics

Up to our humble knowledge, beside time and space our universe consists of energy and matter. As matters mostly take a form of materials, it then became a human's natural urge to study how these materials behave and how this behaviour relates with energy in the framework of space and space-time.

Classical concept that is already well accepted and acknowledged is that the matter or materials take three distinct forms or phases: solid, liquid and gas, with the latter two were put together as fluid. From this concept emerged then the study of material motion via classical mechanics.

However classical mechanics people followed later the idea of Lagrange. In his book, "mechanique analytique' published in the year 1788 he used the energy formulation to give another form of classical or Newtonian mechanics. He did not really care what form or phase the matter takes: for them the matter is represented by solely a mass point and rigid body, in which the matter manifests itself only through its inertia, i.e. through its resistance against translational and rotational motion. It is very centrally useful yet abstract concepts. Modern physics likewise turn its back, according to Professor Truesdell and Noll (1992), one of the founders of Rational Mechanics, i.e. study of how materials behave, putting majority of its interest only on the small particles of matter, rejecting the challenge or question of how the specimen is made up of these particles will behave in typical circumstances in which we meet it. All these materials are deformable. The task now is to describe their mechanical behaviour under all types of well-defined loading circumstances by taking into consideration their deformability and represent a definite physically sound principles they obey. The challenge of taking this task is accepted by the continuum theory of material behaviour or short continuum theory. At the infant stage of this development, the continuum theoretician had their confidence boosted by the popularity and success of Albert Einstein's Relativity theory (1916) utilizing differential geometry and tensor analysis to describe the behaviour of space-time universe. For theoretical continuum mechanician, it is then the material behaviour or behaviour of matter. Since in this theory the materials are embedded one-to-one in three-dimensional Euclidean space and space is axiomatically considered to be continued, i.e. without any defect or void, therefore came the term continuum mechanics. Thus for them the word continuum is suitable for both space-time universe and deformable materials.

As the current technological development of new advanced materials, such as nanomaterials, graphene, functional graded material, functional and intelligent materials, moves forward at rapid phase, the discussion in the continuum theory will never find its end. There are still a lot of open questions to be answered. Even some physically sound principles postulating the behaviour of materials are still needed to be revised. This is a new challenge. We however will focus our discussion in this small booklet only on conventional behaviour of engineering materials. Elasticity, plasticity, viscous materials and fluidity and hyperelasticity are amongst common material behaviours to be named.

The theory of linear elasticity is very old theory and already existed since more than 300 years ago and even about 10 years earlier than the publication of a book Philosophiae Naturalis Principia Mathematica authored by Sir Isaac Newton (1687). In the year 1676, Robert Hooke claimed that he found a law that relates to the property of materials and stated this law in a Latin anagram "ceiiinossssttuu". Two years later 1678 he gave the solution to this anagram and therefore published the law which states that the force produced is related to the extension of the material, i.e. ut tensio sic vis. He himself then called this law as Hooke's Law.

The pathway of the development of the theory of plasticity started very much later. The motivation behind the development of the theory of plasticity is to understand why and how materials yield under complex loading and why some other materials not. The earliest effort toward finding explanation to this behaviour was made by Tresca in the year 1864. He conducted several experiments on extrusion of metals and found out that metals begin to yield when the maximal shear stress reached certain critical value. This finding was then known as Tresca Yield condition and as shear stress hypothesis. Greater impetus for the development of plasticity theory came after Von Mises in 1931 introduced the so-called Von Mises yield criterion or also known as the maximum distortion criterion for ductile materials such as metals. Preceding Maxwell (1856) and Huber (1904), Von Mises developed a yield criterion that states the yielding of a ductile material starts with reaching the critical value of second invariant of deviatoric stress during loading.

1.2 Terminology and Notation

Before we go greatly into mathematical jongleuring of continuum mechanics we should anticipate the kind of terminology, notations and symbols used in this manuscript. Continuum mechanics as part of physics deals inexhaustly with quantities like density, temperature, displacement, velocity, stress and strain, even damage. In physics, people are more concerned with the question if these quantities are observable and measurable. Then there are basically alike. Mathematically, however, these quantities are of different structure. Density and temperature are scalar functions whereby displacement, velocity and acceleration are vector-valued function or sometimes vector-valued vector function. Stress and strain are classified as tensor of second order endowed with tensor characteristics. These physical quantities mentioned here are *field* quantities, i.e. there are functions of position and time as in case of instationary flow.

In this manuscript one uses symbolic notation for scalars, vectors and tensors of second and higher order. Just for the sake of computational simplification, the indicial notation is used based on the Cartesian (orthonormal)coordinate system. The calculation will be done in general symbolically but whenever necessary it will be also done indicially. The results will be presented in both symbolic and indicial forms.

A scalar quantity will be presented as a small or capital letter of Latin or Greek alphabet in italic format such as p for pressure, T for temperature and ρ for density. For a vector a small letter of Latin alphabet boldfaced format will be used for symbolic notation and the same small letter of Latin alphabet with a subscript for indicial notation, such as **v** or v_i for velocity, **a** or a_j for acceleration. For second-order tensor a capital letter of Latin alphabet in boldfaced format or a capital letter of the same Latin alphabet in italic format with two subscripts, e.g. **E** or E_{ij} for strain tensor. However at times there will be an exception to the rule, for example, the identity tensor will be written symbolically as **I** but indically δ_{ij}, or even stress tensor conventionally **S** will be written in indicial notation as σ_{jk}. As for the higher order tensor, in particular tensor of fourth order describing the material properties, its notation rule will be explained extensively in the relevant topic, such as elasticity theory, plasticity theory, creep mechanics and continuum damage mechanics.

1.3 Understanding Some Basic Concepts

Most common word used to describe a material object in order to relate its interaction with the surrounding environment is *body*. This term is coined in order to describe how the material objects that occur in nature respond to application of external forces or other stimuli from surrounding environment.

In continuum mechanics, the term body $\mathcal{B}$ is understood as a mathematical abstraction of an object occurs in nature and refered as a collection of all material points occupying space. Set of all spatial points occupied by the body $\mathcal{B}$ is the region $\mathcal{R}$ of that configuration. But what is material point? Material point is understood as a small tiny of matter or in short, particle, occupies one particular position at one specific time. Each material point is always associated with one particular position or spatial point but not other way around. A position can be empty but particle must always occupy position. Furthermore, neither two or more particles occupy one position nor two or more positions filled by one particle. As for common understanding, the word

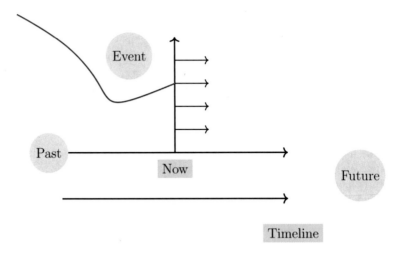

Fig. 1.1 Timeline

material point and particles are synonymous, the same goes to the words spatial point and position.

Figure 1.1 above illustrates graphically the concept of the principle of determinism used as one of the main pillars in the continuum mechanics and rational material theory. As the principle postulates, the the current events are the effect preceded by a cause. And as Laplace formulated it nicely, we can predict the event in the future based on the interaction between cause and effect in the past. So as shown in Fig. 1.1 the concept of future event is non-existent in the continuum mechanics and rational material theory. Ergo the prediction of future event or process is done by looking backward collecting all previous histories and see how these events in the past affect the current or "now" event. In other words, by visualizing the past-event axes with the origin "now" moving monotonically to the right along the axis of timeline, one is able to predict the projection or continuation of past event onto the future.

Let us spend a few words on the term "timeline" in context of the efforts made by Steven Hawking in establishing the Theory of Everything. The direction of timeline as illustrated in the Fig. 1.1 is from the left to the right representing the direction from past to the future. According to Steven Hawking, this is the diction of the Arrow of Time. The following is what he wrote in his book "The Brief History of Time".

> However, there are at least three arrows of time that do distinguish the past from the future. They are the thermodynamic arrow, the direction of time in which disorder increases; the psychological arrow, the direction of time in which we remember the past and not the future; and the cosmological arrow, the direction of time in which the universe expands rather than contracts. I have shown that the psychological arrow is essentially the same as the thermodynamic arrow, so that the two would always point in the same direction.

Clearly the principle of determinism introduced by Laplace and used as one of the pillars of the continuum mechanics is just another form of psychological arrow of time.

Chapter 2
Kinematics

2.1 Lagrangian and Eulerian Approaches

Generally in continuum mechanics there are two different approaches of describing the motion of any material body, or more precisely motion of any material point representing a material body. They are

- Lagrangian approach or body-fixed coordinates system
- Eulerian approach or space-fixed coordinates system.

In the Lagrangian approach, one selects and labels a particular material point among the collection of material points of a body and follows the destiny of this material point throughout its motion process. It puts emphasis on the material description. Eulerian approach however focuses on the spatial description for describing the observed motion. Here is the space point is selected and labelled and the motion is defined as the collection of all material points passing or occupying the selected spatial point at any time.

2.1.1 Reference Versus Current Configuration. Deformation Gradient

One of the important concepts in continuum mechanics is the description of material point and space point. We understand that a material point is a label given to a matter characterized by a mass that occupies any position, $X^i, i = 1, 2, 3$ in 3D Euclidean space at a certain time t. These numbers $X^i, i = 1, 2, 3$ are also known as Lagrangian coordinates of material point. This material point can take any position in space characterized by space point or position $x^k, k = 1, 2, 3$. This space point or coordinate is also known as Eulerian coordinate. There are situations in that position is not occupied by any material point but not the other way round, i.e. one material

N. A. N. Mohamed, *Introduction to Continuum Mechanics for Engineers*,
https://doi.org/10.1007/978-981-99-0811-0_2

point occupies only one space point or position. Furthermore, it is well accepted that neither two or more material points occupy one position nor two or more positions occupy by one material point. Thus the relation between X^i and x^k is bijective or one-to-one. In the Lagrangian perspective, we follow material point as it moves through space. Mathematically, at any current position vector, we have

$$x^k = x^k(X^i, t) \longleftrightarrow \mathbf{x} = \mathbf{x}(\mathbf{X}, t) \tag{2.1}$$

with $\underline{\mathbf{X}}$ is the position vector of the material point at a reference or initial time t_0. So we have specifically

$$X^i = x^i(t = t_0) \longleftrightarrow \mathbf{X} = \mathbf{x}(t = t_0)$$

In Fig. 2.1, the point $A \in \mathcal{B}$, $\mathcal{B}$ is the body, representing a material point occupies the position X^i, $(\mathbf{X})$ at the time t_0 or $t = 0$. The body $\mathcal{B}$ is said to occupy the reference region or conventionally the reference configuration. In this Fig. 2.1, this reference configuration is for sake of simplicity, a circle. After experiencing loading the material point $A \in \mathcal{B}$ moves to other position x_k, $(\underline{\mathbf{x}})$, and still within the body $\mathcal{B}$ but at the current configuration, here in Fig. 2.1, an ellipse. The motion of point A from reference configuration to current configuration is labelled by the displacement vector u^j or $\mathbf{u}$. Thus

$$u^k = x^k - X^k \longleftrightarrow \mathbf{u} = \mathbf{x} - \mathbf{X}$$

Now from Eq. (2.1), the infinitesimal change of the position dx^k is given as

$$dx^k = \frac{\partial x^k(X^i, t)}{\partial X^j} dX^j \longleftrightarrow d\mathbf{x} = \frac{\partial \mathbf{x}(\mathbf{X}, t)}{\partial \mathbf{X}} \cdot d\mathbf{x} \tag{2.2}$$

The expression

$$\frac{\partial x^k(X^i, t)}{\partial X^j} = F_{ij} \longleftrightarrow \frac{\partial \mathbf{x}(\mathbf{X}, t)}{\partial \mathbf{X}} = \mathbf{F}$$

in Eq. (2.2) is called the deformation gradient $\mathbf{F}$ and it describes the deformation of current configuration from the initial reference configuration.

We assume that there is a homeomorphism, between the neighbourhood of reference configuration and the neighbourhood of current configuration, that is, the neighbourhood of a particular material point in the reference configuration is considered smooth, mathematically continuously differentiable. This should be also valid for the neighbourhood of the same material point in the current configuration. Then the relation

$$x^k = x^k(X^i, t) \longleftrightarrow \mathbf{x} = \mathbf{x}(\mathbf{X}, t)$$

is invertible, therefore

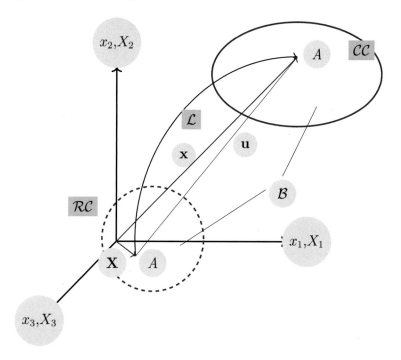

Fig. 2.1 Motion of material point P along the path $\mathcal{L}$. RC: Reference configuration, CC: Current configuration

$$X^k = X^k(x^i, t) \longleftrightarrow \mathbf{X} = \mathbf{X}(\mathbf{x}, t)$$

that is when the Jacobi determinant J is always positive definite.

$$J = \det\left(\frac{\partial x^k(X^i, t)}{\partial X^j}\right) > 0$$

Only if the Jacobi determinant of the deformation gradient is positive definite, the relation between the material coordinates and spatial coordinates described in Eq. (2.1) is called "motion".

2.1.2 Material and Spatial Description of Property

Using these two approaches, i.e. material or Lagrangian approach and spatial or Eulerian approach, any property Φ of a continuum can be described as follows, (a) material or Lagrangian description of Φ

$$\Phi = \Phi(X^i, t)$$

or (b) spatial or Eulerian description of Φ

$$\Phi = \Phi(x^k, t)$$

Thereby Φ can be any quantity of a scalar, a vector or a tensor property. Temperature, velocity or strain field are a few examples to mention.

It becomes much more interesting when we are concerning about the rate of change of this property with respect to time t. For the Lagrangian description of the property Φ, as

$$\Phi = \Phi(X^k, t)$$

with the Lagrangian coordinates, X^k, are fixed, the rate of change of the property Φ with respect to time, also known as material time derivative, is merely the partial differentiation of Φ with respect to time,

$$\frac{d\Phi}{dt} = \frac{\partial \Phi(X^k, t)}{\partial t}\Big|_{X^k} \tag{2.3}$$

The change of temperature described by Eq. (2.3) is measured by an observer with thermometer attached to him moving together with the selected particle or material point x^k.

Imagine a situation where you had fever and changing your location from one place to another place did not cure your fever. However in case the property Φ is described using the spatial Eulerian coordinates, then the material derivative will consist of both local time and convective time derivatives. The following equation illustrates these two time derivatives.

$$\frac{d\Phi}{dt} = \frac{\partial \Phi(x^k, t)}{\partial t}\Big|_{x^k} + \frac{\partial \Phi(x^i, t)}{\partial x^k}\frac{dx^k}{dt} \tag{2.4}$$

The convective time derivative describes the changes in the property Φ as a result of the particle's motion. In symbolical notation Eq. (2.4) can be written as

$$\frac{d\Phi}{dt} = \frac{\partial \Phi}{\partial t}\Big|_{\underline{x}} + \underline{v} \cdot \nabla \Phi \tag{2.5}$$

with

$$\underline{v} = \frac{dx^k}{dt}$$

and the nabla operator $\nabla(.)$

$$\nabla(.) = \frac{\partial(.)}{\partial \underline{x}}$$

2.1.3 Examples and Problems

Example 2.1.1

Examine if the following relation between the Lagrangian and Eulerian coordinates represents the motion.

$$x^1 = X^1 e^{(t)} + X^3(e^{(t)} - 1),\ x^2 = X^2 + X^3(e^{(t)} - e^{(-t)}),\ x^3 = X^3$$

Solution

The solution steps are (i) finding the deformation gradient $\underline{F} = \frac{\partial x^k}{\partial X^l}$ and (ii) check if the Jacobi determinant J is positive definite.
From the relation above we have

$$\frac{\partial x^1}{\partial X^1} = e^{(t)};\ \frac{\partial x^1}{\partial X^2} = 0;\ \frac{\partial x^1}{\partial X^3} = e^{(t)} - 1$$

$$etc.$$

altogether we have then the matrix of deformation gradient

$$(\mathbf{F}) = \left(\frac{\partial x^p}{\partial X^q} \right) = \begin{pmatrix} e^{(t)} & 0 & e^{(t)} - 1 \\ 0 & 1 & (e^{(t)} - e^{(-t)}) \\ 0 & 0 & 1 \end{pmatrix}$$

Therefore the Jacobi determinant $J = det((\underline{F}))$

$$(\mathbf{F}) = det \begin{pmatrix} e^{(t)} & 0 & e^{(t)} - 1 \\ 0 & 1 & (e^{(t)} - e^{(-t)}) \\ 0 & 0 & 1 \end{pmatrix} = e^{(t)}$$

which is always positive for all time t. So the relation given above represents a motion.

Example 2.1.2

Given is the temperature field described using the spatial, Eulerian coordinates as follows:

$$T = T_0(x^1 + x^2)$$

The plane motion of the continuum is characterized by the relation

$$x^1 = X^1 + AtX^2, x^2 = X^2, x^3 = X^3$$

T_0 and A are given constants.

1. Express this temperature field in term of Lagrangian coordinates.
2. Determine the material derivative of this temperature field in both coordinate systems.

Solution

(a) A direct substitution of Lagrangian coordinates into the equation for the given temperature field yields the following equation

$$T = T_0(X^1 + AtX^2 + X^2) = T_0(X^1 + (At + 1)X^2)$$

(b) Material derivative in Lagrangian formalism

$$\frac{DT}{Dt} = \frac{\partial T(X^k, t)}{\partial t}\Big|_{X^k} = AT_0X^2 = AT_0x^2$$

Material derivative in Eulerian formalism

$$\frac{DT}{Dt} = \frac{\partial T(x^k, t)}{\partial t}\Big|_{x^k} + \frac{\partial T(x^i, t)}{\partial x^k}\frac{dx^k}{dt}$$

From the given plane motion we have

$$\frac{dx^1}{dt} = AT_0X^2 = AT_0x^2$$

$$\frac{dx^2}{dt} = 0$$

and

$$\frac{dx^3}{dt} = 0$$

$$\frac{\partial T(x^k, t)}{\partial t}\Big|_{x^k} = 0$$

and furthermore

$$\left(\frac{\partial T(x^i, t)}{\partial x^k}\right) = \begin{pmatrix} T_0 \\ T_0 \\ 0 \end{pmatrix}$$

Thus the material derivative in Eulerian formalism is

$$\frac{DT}{Dt} = 0 + (Ax^2, 0, 0) \begin{pmatrix} T_0 \\ T_0 \\ 0 \end{pmatrix} = AT_0x^2$$

Both calculations lead to the same result however in Lagrangian formalism it is much simpler and straight forward.

Problem 2.1.1

Given is a scalar field $\phi = \phi_0(x^1 - 2x^2)$ represented by spatial, Eulerian coordinates. The motion of the continuum associated with the above field is given as follows:

$$x^1 = X^1 + t^2X^2, x^2 = -t^2X^1 + X^2, x^3 = X^3$$

Find the material derivative of the scalar field ϕ in both Lagrangian and Eulerian formalisms. Compare the results.

Problem 2.1.2

One particular motion is characterized by the following relation between material and spatial coordinates

$$x^1 = X^1 + \alpha X^2, x^2 = X^2, x^3 = X^3$$

Determine the associated deformation gradient, the Jacobian, the metric tensor and Lagrangian strain tensor. Discuss the special cases of $\alpha = 0$ and $\alpha = 1$.

2.2 Lagrangian and Eulerian Displacement Gradient

Any arbitrary body $\mathcal{B}$ will occupy a specific region or configuration in Euclidean three-dimensional space at one specific time τ_o. Now under the influence of external loading, either in form of external forces or external torques, the same body $\mathcal{B}$ will undergo deformation (including translation and rotation). This process of deformation continues with time t. Therefore any material point $A \in \mathcal{B}$ in reference configuration at a time τ_0 will move accordingly to a new position in current configuration of $\mathcal{B}$ at the time t, as shown in Fig. 2.2. The vector $\mathbf{u}$ describing the displacement of the material point or particle A in reference configuration of $\mathcal{B}$ to a new position is called *displacement vector*. This displacement vector $\mathbf{u}$ can be expressed both in Lagrangian

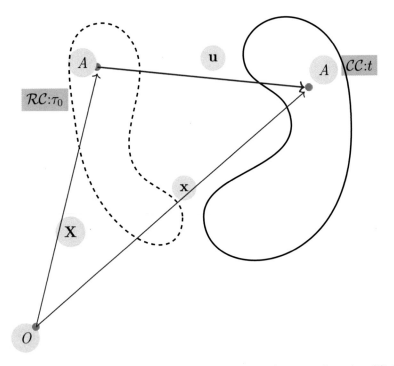

Fig. 2.2 Deformation gradient and displacement gradient. RC: Reference configuration, CC: Current configuration

coordinates and also in Eulerian coordinates. Symbolically the displacement vector is expressed as

$$\mathbf{u} = \mathbf{x} - \mathbf{X}$$

However if it is expressed in Lagrangian, the above equation becomes

$$u_i = x_i(X_j, t) - X_i(t) = u_i(X_j, t)$$

or in Eulerian

$$u_i = x_i(t) - X_i(x_j, t) = u_i(x_j, t)$$

Now if we want to study how is the displacement field varies from one particle to other particle or how this displacement field varies from one position to another position, then we have to use the displacement gradient in Lagrangian or Eulerian coordinates, respectively. Symbollically, the displacement gradient take the following form

$$\mathbf{H} = \nabla \mathbf{u} \tag{2.6}$$

or

$$\mathbf{H} = \nabla(\mathbf{x} - \mathbf{X}) = \nabla\mathbf{x} - \nabla\mathbf{X}$$

with the gradient operator $\nabla(.)$ defined as

$$\nabla(.) := \frac{\partial(.)}{\partial \mathbf{X}}$$

for Lagrangian coordinates and

$$\nabla_x(.) := \frac{\partial(.)}{\partial \mathbf{x}}$$

for Eulerian approach. Now in Lagrangian coordinates the displacement gradient $\mathbf{H}$ take the form

$$\mathbf{H} = \frac{\partial(\mathbf{x} - \mathbf{X})}{\partial \mathbf{X}} = \frac{\partial(\mathbf{x})}{\partial \mathbf{X}} - \frac{\partial(\mathbf{X})}{\partial \mathbf{X}}$$

with

$$\frac{\partial(\mathbf{x})}{\partial \mathbf{X}} = \mathbf{F}$$

the deformation gradient, the displacement gradient $\mathbf{H}$ in Lagrangian coordinates is given as follows:

$$\mathbf{H} = \mathbf{F} - \mathbf{I} \tag{2.7}$$

In Eulerian coordinates, the displacement gradient is defined as follows:

$$\tilde{\mathbf{H}} = \nabla_x \mathbf{u} \tag{2.8}$$

$$\tilde{\mathbf{H}} = \frac{\partial(\mathbf{x} - \mathbf{X})}{\partial \mathbf{x}} = \frac{\partial(\mathbf{x})}{\partial \mathbf{x}} - \frac{\partial(\mathbf{X})}{\partial \mathbf{x}}$$

With

$$\frac{\partial(\mathbf{X})}{\partial \mathbf{x}} = \mathbf{F}^{-1}$$

the inverse of deformation gradient, we get the Eulerian displacement gradient as written below

$$\tilde{\mathbf{H}} = \mathbf{I} - \mathbf{F}^{-1} \tag{2.9}$$

The deformation gradient $\mathbf{F}$ and the associated displacement gradients $\mathbf{H}$ or $\tilde{\mathbf{H}}$ play vital important role in describing the deformation of a body in space. It characterizes the three basic components of motion, translation, rotation and deformation

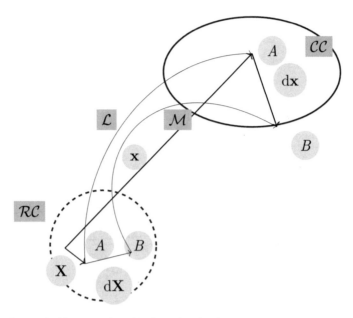

Fig. 2.3 Geometrical Interpretation of Deformation Gradient

(elangation or compression). This property can be easily illustrated in the following Fig. 2.3. The points A and B are two neighbouring material points in a reference configuration $\mathcal{RC}$ at a time $t = 0$. Let the infinitesimal distance vector between these material points in $\mathcal{RC}$ be $d\mathbf{X}$. In the process of deformation, these two points move independently along the paths $\mathcal{L}$ and $\mathcal{M}$, respectively, and take new positions in current configuration $\mathcal{CC}$ at the time t. Now the infinitesimal distance vector between these two points in $\mathcal{CC}$ becomes $d\mathbf{x}$. With the relationship between Lagrangian or material and Eulerian or spatial coordinates given as follows:

$$\mathbf{x} = \mathbf{x}(\mathbf{X})$$

or

$$x_i = x_i(X_j)$$

we have then

$$d\mathbf{x} = \frac{\partial \mathbf{x}}{\partial \mathbf{X}} \cdot d\mathbf{X}$$

The expression $\partial \mathbf{x}/\partial \mathbf{X}$ is the deformation gradient $\mathbf{F}$. The above equation can also be rewritten as follows:

$$d\mathbf{x} = \mathbf{F} \cdot d\mathbf{X}$$

It is clearly from vector or matrix algebra that the above equation describes vector transform from $d\mathbf{X}$ to $d\mathbf{x}$ by $\mathbf{F}$.

2.2.1 Examples and Problems

Example 2.2.1

St. Venant torsion is described by small rotation of the beam's cross-sectional area and the warping of that area in the axial direction. The warping of the cross-sectional area is independent of axial axis X_3 but completely a function of cross-sectional coordinates X_1, X_2 or r, $\theta(X_3)$, with $\theta(X_3)$ is the cross-sectional twist angle per length of the beam. Based on reference configuration, the "motion" St. Venant torsion is described by the following relation.

$$x_1 = X_1 - \frac{\theta}{l} X_2 X_3$$
$$x_2 = X_2 + \frac{\theta}{l} X_3 X_1$$
$$x_3 = X_3 + \psi(X_1, X_2)$$

Whereby l is the length of the beam and $\psi(X_1, X_2)$ is the warping function representing the warping of the beam's cross section. Find the displacement vector $\mathbf{u}$. Determine the deformation gradient $\mathbf{F}$ and the corresponding Lagrangian displacement gradient $\mathbf{H}$. Determine also the Eulerian displacement gradient $\bar{\mathbf{H}}$ (Fig. 2.4).

Solution

From the equations relating the St. Venant torsion in both material and spatial coordinates, X_i and x_j, $i, i = 1, 2, 3$

$$x_1 = X_1 - \frac{\theta}{l} X_2 X_3$$
$$x_2 = X_2 + \frac{\theta}{l} X_3 X_1$$
$$x_3 = X_3 + \psi(X_1, X_2)$$

The displacement vector u_i can be obtained directly using the equation

$$u_i = x_i - X_i$$

which produces

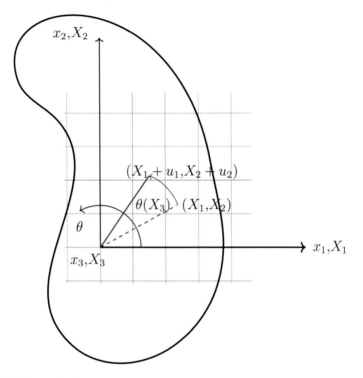

Fig. 2.4 St. Venant Torsion

$$u_1 = -\frac{\theta}{l} X_2 X_3$$

$$u_2 = \frac{\theta}{l} X_3 X_1$$

$$u_3 = \psi(X_1, X_2)$$

the components of the displacement vector.

We have the components of the deformation gradient, $\partial x_i / \partial X_j$ by differentiating x_i with respect to material coordinates X_j, as follows:

$$\partial x_1 / \partial X_1 = 1$$

$$\partial x_1 / \partial X_2 = -\frac{\theta}{l} X_3$$

$$\partial x_1 / \partial X_2 = -\frac{\theta}{l} X_2$$

$$\partial x_2 / \partial X_1 = \frac{\theta}{l} X_3$$

$$\partial x_2 / \partial X_2 = 1$$

$$\partial x_2/\partial X_3 = \frac{\theta}{l} X_1$$
$$\partial x_3/\partial X_1 = \partial \psi(X_1, X_2)/\partial X_1$$
$$\partial x_3/\partial X_2 = \partial \psi(X_1, X_2)/\partial X_2$$
$$\partial x_3/\partial X_3 = 1$$

Rearranging the equations above, we have then the deformation gradient **F**

$$(\mathbf{F}) = \partial x_i/\partial X_j = \begin{pmatrix} 1 & -(\theta/l)X_3 & -(\theta/l)X_2 \\ (\theta/l)X_3 & 1 & (\theta/l)X_1 \\ \partial \psi(X_1, X_2)/\partial X_1 & \partial \psi(X_1, X_2)/\partial X_2 & 1 \end{pmatrix}$$

Now to find the displacement gradient we use Eq. (2.7) and get

$$(\mathbf{H}) = \partial u_i/\partial X_j = \begin{pmatrix} 0 & -(\theta/l)X_3 & -(\theta/l)X_2 \\ (\theta/l)X_3 & 0 & (\theta/l)X_1 \\ \partial \psi(X_1, X_2)/\partial X_1 & \partial \psi(X_1, X_2)/\partial X_2 & 0 \end{pmatrix}$$

Of course we can also obtain the displacement gradient from the calculated displacement vector **u**.

Problem 2.2.1

Given is the following motion

$$x_1 = X_1$$
$$x_2 = X_1 + X_2$$
$$x_3 = X_1 + X_2 + X_3$$

Verify that the above equations representing a motion. Calculate the displacement gradient in both approaches, Lagrangian and Eulerian.

Problem 2.2.2

Given is a body with the motion

$$x_1 = 1 + X_1^2 X_2$$
$$x_2 = 1 + X_2^2 X_1$$
$$x_3 = X_3$$

Perform the task as in the problem (2.2.1). Verify that the above equations represent a motion. Calculate the displacement gradient in both approaches, Lagrangian and Eulerian.

2.3 Strain Tensor

Deformation gradient is in fact very useful tool in continuum mechanics to describe a motion of an object. The only problem is, it also covers the motion of rigid bodies, which are translation and rotation. So there is a need to find a measure to describe the deformation of non-rigid bodies, which is more real-world prevalent. Starting from the definition of rigid body, which states that if any two material points in a body do not change their distance from each other under loading, then the body is said to be rigid, otherwise the body is considered deformable body. Deformable bodies are bodies that are able to undergo deformation. For the case of one-dimensional continuum, the measure of deformation is simply the percentage of the extension for the tensile loading (or shortening for compression) of the body from the original length. This measure is conventionally called strain. Similar to the concept of defining strain measure in one-dimensional case, we use the difference of the quadrature of the distance between any two material points infinitesimally closed to each other both in the current and reference configurations to define the strain measure for three-dimensional continuum. The quadrature of the infinitesimal distances is purposely to capture the three dimensionality of the observed continuum.

$$\Delta^2 := ds^2 - dS^2 \tag{2.10}$$

ds and dS are the length of line element connecting these two arbitrary material points in close to each other in the current and reference configuration respectively. With

$$ds^2 = d\mathbf{x} \cdot d\mathbf{x}$$

and

$$dS^2 = d\mathbf{X} \cdot d\mathbf{X}$$

We can express this measure in two ways, either in material coordinates or in spatial coordinates. From Sect. 2.2, Eq. (2.5), we know that both coordinates are related, that is

$$d\mathbf{x} = \mathbf{F} \cdot d\mathbf{X} \tag{2.11}$$

or in index notation

$$dx^i = \frac{\partial x^i(X^k, t)}{\partial X^j} dX_j = F^{ij} dX_j$$

Now the[1] difference between the quadratures of infinitesimal distance between two material points in current configuration and reference configuration is given as

$$\Delta^2 := ds^2 - dS^2$$

$$d\mathbf{x} \cdot d\mathbf{x} - d\mathbf{X} \cdot d\mathbf{X}$$

and with Eq. (2.11) the above equation can be written as

$$\Delta^2 = (\mathbf{F} \cdot d\mathbf{X}) \cdot (\mathbf{F} \cdot d\mathbf{X}) - d\mathbf{X} \cdot d\mathbf{X}$$

or

$$\Delta^2 = d\mathbf{X} \cdot (\mathbf{F}^\mathbf{T} \cdot \mathbf{F}) \cdot d\mathbf{X} - d\mathbf{X} \cdot (\mathbf{I}) \cdot d\mathbf{X}$$

After vectorial factorization we have

$$\Delta^2 = d\mathbf{X} \cdot (\mathbf{F}^\mathbf{T} \cdot \mathbf{F} - \mathbf{I}) \cdot d\mathbf{X}$$

Now in this equation, we have a measure for deformation constructed based on material or Lagrange coordinates d$\mathbf{X}$. We define this measure as Lagrangian strain tensor $\mathbf{E}$.

$$\mathbf{E} = \frac{1}{2}(\mathbf{F}^\mathbf{T} \cdot \mathbf{F} - \mathbf{I}) \tag{2.12}$$

So the difference of the quadratures of the distance of any two material points in current and reference configuration can be expressed by a strain tensor $\mathbf{E}$ as

$$\Delta^2 = 2\, d\mathbf{X} \cdot \mathbf{E} \cdot d\mathbf{X}$$

In term of spatial or Eulerian coordinates, the corresponding strain tensor have the following form

$$\Delta^2 = d\mathbf{x} \cdot (\mathbf{I} - \mathbf{F}^{-\mathbf{T}} \cdot \mathbf{F}^{-1}) \cdot d\mathbf{x}$$

or

$$\Delta^2 = 2\, d\mathbf{x} \cdot \mathbf{E} \cdot d\mathbf{x}$$

With the Eulerian strain tensor $\mathbf{e}$ defined as

$$\mathbf{e} = \frac{1}{2}(\mathbf{I} - \mathbf{F}^{-\mathbf{T}} \cdot \mathbf{F}^{-1}) \tag{2.13}$$

---

[1] Note that the index for the material and spatial coordinates are written in two different ways. One is in superscript and the another one is in subscript. This convention has to do with the basis vectors of the tensor used.

In indicial notation, these two strain tensors take the following form

$$\epsilon_{ij} = \frac{1}{2}\left(\frac{\partial x^i(X^p,t)}{\partial X^k}\frac{\partial x^j(X^p,t)}{\partial X^k} - \delta_{ij}\right) \tag{2.14}$$

$$e_{ij} = \frac{1}{2}\left(\delta_{ij} - \frac{\partial X^i(x^p,t)}{\partial x^k}\frac{\partial X^j(x^p,t)}{\partial x^k}\right) \tag{2.15}$$

One notices that these two strain tensors are nonlinear strain measures and are needed to be modified for smaller deformation. This nonlinearity is often called geometrical nonlinearity which cannot be ignored for finite deformation, such as deformation of rubber or any material with rubber-like elasticity. Lagrangian strain tensor is important for describing the deformation of solid materials while Eulerian counterpart is commonly used in describing fluid flow.

In kinematics of deformation, the tensor $\mathbf{F}^T \cdot \mathbf{F} = \mathbf{C}$ is called "right Cauchy–Green tensor" while $\mathbf{F} \cdot \mathbf{F}^T = \mathbf{B}$ is called "left Cauchy–Green" tensor. Using these tensor, we can simplify Eqs. (2.12) and (2.13) to become

$$\mathbf{E} = \frac{1}{2}(\mathbf{C} - \mathbf{I})$$

and

$$\mathbf{e} = \frac{1}{2}(\mathbf{I} - \mathbf{B}^{-T})$$

Linearization of both strain tensors can be accomplished by utilizing the displacement gradient $\mathbf{H}$ or $\tilde{\mathbf{H}}$. From the equation relating displacement and deformation gradient, we rewrite $\mathbf{F}$ in terms of $\mathbf{H}$, we have

$$\mathbf{F} = \mathbf{H} + \mathbf{I}$$

Now the Lagrangian strain tensor $\mathbf{E}$ can be written as

$$\mathbf{E} = \frac{1}{2}((\mathbf{H} + \mathbf{I})^T \cdot (\mathbf{H} + \mathbf{I}) - \mathbf{I})$$

Upon expansion we have

$$\mathbf{E} = \frac{1}{2}(\mathbf{H} + \mathbf{H}^T + \mathbf{H}^T \cdot \mathbf{H}) \tag{2.16}$$

Expressed in indicial notation the Lagrangian strain tensor takes the following form as derivatives of displacement field of the body, i.e.

$$\epsilon^{ij} = \frac{1}{2}\left(\frac{\partial u^i}{\partial X^j} + \frac{\partial u^j}{\partial X^i} + \frac{\partial u^j}{\partial X^k}\frac{\partial u^i}{\partial X^k}\right)$$

Now the geometrical linear strain tensor, $\hat{\mathbf{E}}$ is obtained simply by omitting the second-order terms $\mathbf{H}^T \cdot \mathbf{H}$ in Eq. (2.16). We have then

$$\hat{\mathbf{E}} = \frac{1}{2}(\mathbf{H} + \mathbf{H}^T) \tag{2.17}$$

$$\hat{\epsilon}^{ij} = \frac{1}{2}\left(\frac{\partial u^i}{\partial X^j} + \frac{\partial u^j}{\partial X^i}\right)$$

This is the linear version of Lagrangian strain tensor and has a wide range of applications in the elasticity of small deformation. In German literature, it is also known as deformator. The reader is also advised to determine herself the linear version of Eulerian strain tensor as to obtain the complete picture of linear strain measures available.

2.3.1 Examples and Problems

Example 2.3.1

We refer back to the St. Venant Torsion discussed in previous example, where relation of both material and spatial coordinates is given as follows:

$$x_1 = X_1 - \frac{\theta}{l}X_2X_3$$

$$x_2 = X_2 + \frac{\theta}{l}X_3X_1$$

$$x_3 = X_3 + \psi(X_1, X_2)$$

The displacement vector u_i can be obtained directly using the equation

$$u_i = x_i - X_i$$

which produces

$$u_1 = -\frac{\theta}{l}X_2X_3$$

$$u_2 = \frac{\theta}{l}X_3X_1$$

$$u_3 = \psi(X_1, X_2)$$

the components of the displacement vector. From here we can easily obtain the linearized Lagrangian strain tensor, whose components are

$$\partial u_1/\partial X_1 = 0$$

$$\partial u_1/\partial X_2 = -\frac{\theta}{l}X_3$$

$$\partial u_1/\partial X_3 = -\frac{\theta}{l}X_2$$

$$\partial u_2/\partial X_1 = -\frac{\theta}{l}X_3$$

$$\partial u_2/\partial X_2 = 0$$

$$\partial u_2/\partial X_3 = \frac{\theta}{l}X_1$$

$$\partial u_3/\partial X_1 = \partial \psi(X_1, X_2)/\partial X_1$$

$$\partial u_3/\partial X_2 = \partial \psi(X_1, X_2)/\partial X_2$$

$$\partial u_3/\partial X_3 = 0$$

Rearranging the components to get the displacement gradient and its transpose, we have then

$$(\mathbf{H}) = \partial \mathbf{u_i}/\partial \mathbf{X_j} = \begin{pmatrix} 0 & -(\theta/l)X_3 & -(\theta/l)X_2 \\ (\theta/l)X_3 & 0 & (\theta/l)X_1 \\ \partial \psi(X_1, X_2)/\partial X_1 & \partial \psi(X_1, X_2)/\partial X_2 & 0 \end{pmatrix}$$

and

$$\mathbf{H} + \mathbf{H^T} = \begin{pmatrix} 0 & 0 & \partial \psi/\partial X_1 - (\theta/l)X_2 \\ 0 & 0 & \partial \psi/\partial X_2 + (\theta/l)X_1 \\ \partial \psi/\partial X_1 - (\theta/l)X_2 & \partial \psi/\partial X_2 + (\theta/l)X_1 & 0 \end{pmatrix}$$

Therefore, the linearized Lagrangian strain has the following form

$$(\mathbf{\hat{E}}) = \frac{1}{2} \begin{pmatrix} 0 & 0 & \partial \psi/\partial X_1 - (\theta/l)X_2 \\ 0 & 0 & \partial \psi/\partial X_2 + (\theta/l)X_1 \\ \partial \psi/\partial X_1 - (\theta/l)X_2 & \partial \psi/\partial X_2 + (\theta/l)X_1 & 0 \end{pmatrix}$$

we note that the Lagrangian strain tensor does not depend on the X_3 coordinate, i.e. the beam's axis. It solely depends on the cross-sectional coordinates, X_1 and X_2. It tells us, that the deformation of the beam during the St. Venant torsion occurs only to the cross section of the beam and it is the same along the beam axis. The deformation that we have here is the rotation of the cross section as well the warping of the cross-sectional area.

Problem 2.3.1

Verify that the transformation below represents a motion

$$x_1 = X_1 e^t + X_3(e^t - 1)$$

$$x_2 = X_2 e^{-t} + X_3(e^t - e^{-t})$$

$$x_3 = X_3$$

If it is a motion, find the deformation gradient $\mathbf{F}$. Determine the corresponding Lagrangian and Eulerian strain tensor.

Problem 2.3.2

As a result of displacement, the particles (X_1, X_2, X_3) of a body are situated at the points with the coordinates

$$x_1 = X_1 + \epsilon X_1, x_2 = X_2, x_3 = X_3$$

(ϵ = const) with reference to a spatial Cartesian coordinate system x_i. What happens to the material lines with both ends are initially at $(0, 0, 0)$ and $(1, 0, 0)$ and $0, 0, 0)$ and $(0, 1, 0)$ respectively following the motion described by the equations above? Consider the two cases: $\epsilon > 0$ and $|1 < \epsilon < 0?$. Explain the results. Determine the corresponding lagrangian and Eulerian strain tensors of this motion.

Problem 2.3.3

Given is the following motion

$$x_1 = X_1 \cos \frac{\pi}{4} t + X_2 \sin \frac{\pi}{4} t$$

$$x_2 = -X_1 \sin \frac{\pi}{4} t + X_2 \cos \frac{\pi}{4} t$$

$$x_3 = X_3$$

1. Find the position of a material line represented by a pair of cartesian coordinates A(0, 1, 0), B(0, 2, 0) initially, after the time $t = 1$ lapses.
2. Calculate the length of this material line before the motion and after $t = 1$ motion.
3. Calculate the deformation gradient $\mathbf{F}$, the right and left Green tensors, $\mathbf{C}$, $\mathbf{B}$ and
4. Show that the Lagrangian and Eulerian strain tensors, $\mathbf{E}$, $\mathbf{e}$, for this motion are null tensors.

Explain the results.

Problem 2.3.4

For both homogeneous deformations described below

$$x_1 = X_1 + \kappa X_2; \ \ x_2 = X_2; \ \ x_3 = X_3$$

and

$$x_1 = \psi X_1; \ \ x_2 = \frac{1}{\psi} X_2; \ \ x_3 = X_3$$

obtain the corresponding deformation gradients. Show that they have the same strain tensor if $\kappa = |\psi - \frac{1}{\psi}|$.

2.4   Polar Decomposition Theorem

Polar decomposition theorem is very useful tool to describe a rotation and deformation of a body after excluding the translational motion. This theorem states that any positive definite tensor can be decomposed into a scalar multiplication of a symmetry tensor and an proper orthogonal tensor. In continuum mechanics, this theorem is applied solely to deformation gradient as it describes the motion.

$$\mathbf{F} = \mathbf{R} \cdot \mathbf{U} = \mathbf{V} \cdot \mathbf{R} \tag{2.18}$$

The tensors $\mathbf{U} = \mathbf{U}^{\mathrm{T}}$ and $\mathbf{V} = \mathbf{V}^{\mathrm{T}}$ are two symmetrical tensors while $\mathbf{R}$ is a proper orthogonal tensor with the property

$$\mathbf{R} \cdot \mathbf{R}^{\mathrm{T}} = \mathbf{R}^{\mathrm{T}} \cdot \mathbf{R} = \mathbf{I}$$

and

$$\det(\mathbf{R}) = 1$$

In continuum mechanics, the tensors $\mathbf{U}$ and $\mathbf{V}$ are called the right stretch and the left stretch tensors, respectively. Both tensors $\mathbf{U}$ and $\mathbf{V}$ can be obtained from either right Cauchy–Green tensor $\mathbf{C}$ or left Cauchy–Green tensor $\mathbf{B}$. By definition

$$\mathbf{C} = \mathbf{F}^{\mathrm{T}} \cdot \mathbf{F} = (\mathbf{R} \cdot \mathbf{U})^{\mathrm{T}} \cdot \mathbf{R} \cdot \mathbf{U} = \mathbf{U}^{\mathrm{T}} \cdot \mathbf{R}^{\mathrm{T}} \cdot \mathbf{R} \cdot \mathbf{U}$$

Since $\mathbf{U} = \mathbf{U}^{\mathrm{T}}$ and $\mathbf{R}^{\mathrm{T}} \cdot \mathbf{R} = \mathbf{I}$ we have

$$\mathbf{C} = \mathbf{U}^2$$

or

$$\mathbf{U} = \mathbf{C}^{1/2}$$

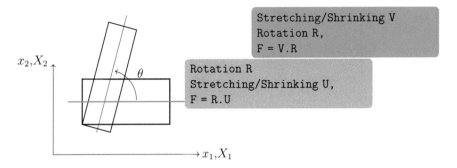

Fig. 2.5 Polar Decomposition

In similar fashion can be done for obtaining the tensor $\mathbf{V}$. Using the left Cauchy–Green tensor $\mathbf{B}$, we have

$$\mathbf{V} = \mathbf{B}^{1/2}$$

The square root of the tensor $\mathbf{C}$ or $\mathbf{B}$ can be obtained using the isotropic tensor function via Cayley–Hamilton theorem or simply finding the eigenvalues of the strain tensors $\mathbf{C}$ or $\mathbf{B}$.

As shown in Fig. 2.5 the deformation of the body, here the blue rectangle to black rectangle happens by either rotation of blue rectangle followed by stretching or shrinking or by stretching or shrinking and then followed by rigid body rotation. This is the reason why the tensor $\mathbf{U}$ or $\mathbf{V}$ are called right or left stretch tensor respectively. The orthogonal tensor $\mathbf{R}$ which describes the rotation of the body can be easily obtained from Eq. (2.18). From either of these equations,

$$\mathbf{F} = \mathbf{R} \cdot \mathbf{U} = \mathbf{V} \cdot \mathbf{R}$$

we have then either

$$\mathbf{R} = \mathbf{F} \cdot \mathbf{U}^{-1}$$

or

$$\mathbf{R} = \mathbf{V}^{-1} \cdot \mathbf{F}$$

From the above equation, we can find a relation between the right stretch and left stretch tensor $\mathbf{U}$ or $\mathbf{V}$,

$$\mathbf{U} = \mathbf{F}^{-1} \cdot \mathbf{V} \cdot \mathbf{F}$$

2.4.1   Examples and Problems

Example 2.4.1

A body is rotated in three-dimensional space by a orthogonal tensor $\mathbf{R}$

$$\mathbf{R} = (1/3) \begin{pmatrix} 1 & 2 & 2 \\ 2 & 0 & -2 \\ -2 & 2 & -1 \end{pmatrix}$$

Verify that the tensor $\mathbf{R}$ is a proper orthogonal tensor and find the axis of rotation and corresponding rotation angle.

Solution

The determinant of $\mathbf{R}$ can be easily calculated and obtained.

$$\det \mathbf{R} = +1$$

Another property of an orthogonal tensor is

$$\mathbf{R} \cdot \mathbf{R}^T = \mathbf{R}^T \cdot \mathbf{R} = \mathbf{I}$$

We use Falk's matrix scheme to verify this property.

$$(1/9) \begin{pmatrix} & & & 1 & 2 & 2 \\ & & & 2 & 0 & -2 \\ & & & -2 & 2 & -1 \\ 1 & 2 & -2 & 1 & 0 & 0 \\ 2 & 0 & 2 & 0 & 1 & 0 \\ 2 & -2 & -1 & 0 & 0 & 1 \end{pmatrix}$$

or

$$(1/9) \begin{pmatrix} & & & 1 & 2 & -2 \\ & & & 2 & 0 & 2 \\ & & & 2 & -2 & -1 \\ 1 & 2 & 2 & 1 & 0 & 0 \\ 2 & 0 & -2 & 0 & 1 & 0 \\ -2 & 2 & -1 & 0 & 0 & 1 \end{pmatrix}$$

Rotation axis can be obtained by determining the eigenvector of $\mathbf{R}$ corresponding to its real eigenvalues. After some calculations the eigenvalues of $\mathbf{R}$ are

$$1, 1/3(-1 + 2i\sqrt{2}), 1/3(-1 - 2i\sqrt{2})$$

Since 1 is the only real eigenvalue so we determine the eigenvector associated with eigenvalue equals 1. Again using symbolic software, we obtain the corresponding eigenvector to the real eigenvalue of $\mathbf{R}$ as follows:

$$\mathbf{v} = (1, 1, 0)$$

To find the rotation angle, we select one arbitrary vector and rotate this vector using the rotation tensor $\mathbf{R}$. The scalar product between the selected vector and the resulted vector with give us the required rotation angle. We select the start vector $\mathbf{a}$ as $\mathbf{a} = (1/\sqrt{3}(1, 1, 1)$. Then this vector is rotated by $\mathbf{R}$ to produce $\mathbf{b}$ according to

$$\mathbf{b} = \mathbf{R} \cdot \mathbf{a} = (1/3)(1/\sqrt{3}) \begin{pmatrix} 1 & 2 & 2 \\ 2 & 0 & -2 \\ -2 & 2 & -1 \end{pmatrix} (1, 1, 1)$$

$$= \frac{1}{9\sqrt{3}}(5, 1, -1)$$

The rotation angle ϕ can be obtained from the scalar product of these two vectors via the following equation for scalar product, which produces

$$\mathbf{a} \cdot \mathbf{b} = \sqrt{(\mathbf{a} \cdot \mathbf{a})}\sqrt{(\mathbf{b} \cdot \mathbf{b})} \cos \phi$$

or

$$\cos \phi = \frac{\mathbf{a} \cdot \mathbf{b}}{\sqrt{(\mathbf{a} \cdot \mathbf{a})}\sqrt{(\mathbf{b} \cdot \mathbf{b})}}$$

With $\mathbf{a} \cdot \mathbf{b} = 5/3$, $\mathbf{a} \cdot \mathbf{a} = 3$ and $\mathbf{b} \cdot \mathbf{b} = 3$ the angle ϕ equals to

$$\phi = \arccos(5/9)$$

Problem 2.4.1

Apply the polar decomposition theorem to the motion described below

$$x_1 = \sqrt{3}X_1; \quad x_2 = 2X_2; \quad x3 = \sqrt{3}X_3 - X_2$$

Problem 2.4.2

Apply the polar decomposition theorem to the shear motion described below

$$x_1 = X_1 + \kappa X_2; \quad x_2 = \kappa X_1 + X_2; \quad x_3 = X_3$$

Problem 2.4.3

Derive the relationship between the "right" and "left" CAUCHY–GREEN-tensor.

2.5   Principal Strains and Directions

As mentioned earlier in Sect. 2.3, the Lagrangian strain tensor $\mathbf{E}$ is a measure of strain suffered by the body during the deformation. It describes the state of strain of a body under loading in three-dimensional Euclidean space. Since the strain progresses in all directions, it is very important to know in which direction the strain of the body takes the highest value under a specific loading. This is particularly useful when we are to measure the strain. The answer of this question is definitely determining the principal strains and their directions or their corresponding principal axes. As the state of strain is represented by Lagrangian strain tensor $\mathbf{E}$ or for small deformation, by the geometrically linear Lagrangian strain tensor $\hat{\mathbf{E}}$ and

$$\hat{\mathbf{E}} = (1/2)(\mathbf{H} + \mathbf{H}^{\mathsf{T}}) \longleftrightarrow \epsilon_{ij} = \epsilon^{ij} = \frac{1}{2} \left(\frac{\partial u^i}{\partial X^j} + \frac{\partial u^j}{\partial X^i} \right)$$

The direct method of finding the principal values and the principal axes of the strain tensor is through solving the eigenvalue problem of the strain tensor.[2] Let $\mathbf{a}$ be any cartesian vector in euclidean space then the eigenvalue problem is to find ϵ that fulfil the following condition

$$\hat{\mathbf{E}} \cdot \mathbf{a} = \epsilon \, \mathbf{a} \longleftrightarrow \epsilon_{ij} a_j = \epsilon \, a_i$$

or

$$(\hat{\mathbf{E}} - \epsilon \, \mathbf{I}) \cdot \mathbf{a} = \mathbf{0} \longleftrightarrow (\epsilon_{ij} - \epsilon \, \delta_{ij}) a_j = 0$$

The condition for non-trivial solution for this eigenvalue problem is

$$\det(\hat{\mathbf{E}} - \epsilon \, \mathbf{I}) = 0 \longleftrightarrow \det(\epsilon_{ij} - \epsilon \, \delta_{ij}) = 0)$$

---

[2] The other method is exploiting the extremal value determination using Lagrange multiplicator method.

The corresponding characteristic equation for the eigenvalues of ϵ is obtained by expanding the above condition, we have then

$$\epsilon^3 - I_1\epsilon^2 + I_2\epsilon - I_3 = 0$$

where I_1, I_2 and I_3 are invariants of the tensor ϵ_{ij} and given as follows.

$$I_1 = tr\,(\epsilon_{ij}) = \epsilon_{11} + \epsilon_{22} + \epsilon_{33}$$

$$I_2 = \det\begin{pmatrix} \epsilon_{11} & \epsilon_{12} \\ \epsilon_{21} & \epsilon_{22} \end{pmatrix} + \det\begin{pmatrix} \epsilon_{22} & \epsilon_{23} \\ \epsilon_{32} & \epsilon_{33} \end{pmatrix} + \det\begin{pmatrix} \epsilon_{11} & \epsilon_{13} \\ \epsilon_{31} & \epsilon_{33} \end{pmatrix}$$

and

$$I_3 = \det(\epsilon_{ij})$$

The principal axis is then the vector **a** associated with each eigenvalue calculated from the characteristics equation above. The other method of finding the extremal values of strain is via orthogonal or rigid body transformation of strain tensor. This method is particularly useful for finding the extremal value of strain in a plane strain problem. Let **Q** be an orthogonal tensor representing a rotation about one specific axis. Then the transformed linearized Lagrange strain tensor $\hat{\mathbf{E}}^*$ is

$$\hat{\mathbf{E}}^* = \mathbf{Q} \cdot \hat{\mathbf{E}} \cdot \mathbf{Q}^T$$

$\hat{\mathbf{E}}^*$ and $\hat{\mathbf{E}}$ are the same linearized Lagrangian strain tensors. The only difference is that they are written based on different basis systems. In indicial notation, the above transformation equation takes the following form

$$\epsilon_{ij}^* = Q_{ip}\epsilon_{pq}Q_{qj}$$

After expansion according to Einstein's convention letting the dummy indicies i, j free, we have

$$\epsilon_{ij}^* = Q_{i1}\epsilon_{11}Q_{1j} + Q_{i2}\epsilon_{21}Q_{1j} + Q_{i3}\epsilon_{31}Q_{1j}$$

$$+Q_{i1}\epsilon_{12}Q_{2j} + Q_{i2}\epsilon_{22}Q_{2j} + Q_{i3}\epsilon_{32}Q_{2j}$$

$$+Q_{i1}\epsilon_{13}Q_{3j} + Q_{i2}\epsilon_{23}Q_{3j} + Q_{i3}\epsilon_{33}Q_{3j}$$

For example for the component ϵ_{11}^*, with $i = 1$, $j = 1$ we get

$$\epsilon_{11}^* = Q_{11}\epsilon_{11}Q_{11} + Q_{12}\epsilon_{21}Q_{11} + Q_{13}\epsilon_{31}Q_{11}$$

$$+Q_{11}\epsilon_{12}Q_{21} + Q_{12}\epsilon_{22}Q_{21} + Q_{13}\epsilon_{32}Q_{21}$$

$$+Q_{11}\epsilon_{13}Q_{31} + Q_{12}\epsilon_{23}Q_{31} + Q_{13}\epsilon_{33}Q_{31}$$

In case of a plane strain with θ rotation angle about (3)-axis,

$$(\hat{E}) = \begin{pmatrix} \epsilon_{11} & \epsilon_{12} & 0 \\ \epsilon_{21} & \epsilon_{22} & 0 \\ 0 & 0 & 0 \end{pmatrix}$$

and the transformation tensor

$$(Q) = \begin{pmatrix} \cos\theta & -\sin\theta & 0 \\ \sin\theta & \cos\theta & 0 \\ 0 & 0 & 1 \end{pmatrix}$$

the above equation is simplified to

$$\epsilon_{11}^* = \epsilon_{11}(\cos\theta)^2 + \epsilon_{22}(\sin\theta)^2 + 2\epsilon_{12}\sin\theta\cos\theta$$

The component ϵ_{11}^* and ϵ_{11}^* can be calculated by the same manner.

2.5.1 Examples and Problems

Example 2.5.1

In an experiment for the determination of plane strain, a 45° strain gauge rosette is used. From the measurement, it is found that

$$\epsilon_{0°}^* = 100\,\mu, \epsilon_{45°}^* = 50\,\mu, \epsilon_{90°}^* = -50\,\mu,$$

Determine all the components of plane strain tensor.

Solution

Use this equation

$$\epsilon_{11}^* = \epsilon_{11}(\cos\theta)^2 + \epsilon_{22}(\sin\theta)^2 + 2\epsilon_{12}\sin\theta\cos\theta$$

for ϵ_{11}^* at three different angles of 0°, 45° and 90° and get

$$\epsilon_{0°}^* = \epsilon_{11} + 0 + 0 = 100$$

$$\epsilon_{90°}^* = 0 + \epsilon_{22} + 0 = -50$$

$$\epsilon^*_{45°} = \epsilon_{11}(1/2) + \epsilon_{22}(1/2) + \epsilon_{12} = 100$$

We have then

$$\epsilon_{11} = 100\,\mu, \epsilon_{22} = -50\,\mu, \epsilon_{12} = 75\,\mu$$

The principal values of this strain can be obtained using Mohr's circle.

Example 2.5.2

If **T** is any tensor of second order, show that it obeys its own characteristic equation

$$\mathbf{T}^3 - I_T\mathbf{T}^2 + II_T\mathbf{T} - III_T\mathbf{I} = 0$$

I_T, I_T and I_T are the first, second and third invariant of **T**.

Solution

Let λ_1, λ_2 and λ_3 are the eigenvalues of the tensor **T**. They satisfies the characteristic equation automatically, i.e.

$$\lambda_1^3 - I_T\lambda_1^2 + II_T\lambda_1 - III_T = 0$$

$$\lambda_2^3 - I_T\lambda_2^2 + II_T\lambda_2 - III_T = 0$$

$$\lambda_3^3 - I_T\lambda_3^2 + II_T\lambda_3 - III_T = 0$$

Rearranging together in a form of a matrix

$$\begin{pmatrix} \lambda_1^3 & 0 & 0 \\ 0 & \lambda_2^3 & 0 \\ 0 & 0 & \lambda_3^3 \end{pmatrix} - I_T \begin{pmatrix} \lambda_1^2 & 0 & 0 \\ 0 & \lambda_2^2 & 0 \\ 0 & 0 & \lambda_3^2 \end{pmatrix} - II_T \begin{pmatrix} \lambda_1 & 0 & 0 \\ 0 & \lambda_2 & 0 \\ 0 & 0 & \lambda_3 \end{pmatrix}$$

$$- III_T \begin{pmatrix} 1 & 0 & 0 \\ 0 & 1 & 0 \\ 0 & 0 & 1 \end{pmatrix} = \begin{pmatrix} 0 & 0 & 0 \\ 0 & 0 & 0 \\ 0 & 0 & 0 \end{pmatrix}$$

One notes that the matrix

$$\begin{pmatrix} \lambda_1 & 0 & 0 \\ 0 & \lambda_2 & 0 \\ 0 & 0 & \lambda_3 \end{pmatrix}$$

is a spectral matrix of the tensor $\mathbf{T}$, denoted as $\hat{\mathbf{T}}$, so that we can rewrite the above matrix-form of characteristic equation as

$$\hat{\mathbf{T}}^3 - I_T\hat{\mathbf{T}}^2 + II_T\hat{\mathbf{T}} - III_T\mathbf{I} = \mathbf{0}$$

A tensor of any order can be transformed to the same tensor based on different bases system via an orthogonal transformation. In this case, a spectral matrix of tensor $\mathbf{T}$ can be transformed back to any other form using an appropriate orthogonal matrix, $\mathbf{Q}$ with the property $\mathbf{Q} \cdot \mathbf{Q}^T = \mathbf{Q}^T \cdot \mathbf{Q} = \mathbf{I}$ and det $\mathbf{Q} = +1$. Now using the orthogonal transformation, we transform the spectral form of the tensor $\mathbf{T}$ to any other form through

$$\mathbf{Q} \cdot \hat{\mathbf{T}} \cdot \mathbf{Q}^T = \mathbf{T}$$

$$\mathbf{Q} \cdot \hat{\mathbf{T}}^2 \cdot \mathbf{Q}^T = (\mathbf{Q} \cdot \hat{\mathbf{T}} \cdot \mathbf{Q}^T) \cdot (\mathbf{Q} \cdot \hat{\mathbf{T}} \cdot \mathbf{Q}^T) = \mathbf{T}^2$$

$$\mathbf{Q} \cdot \hat{\mathbf{T}}^3 \cdot \mathbf{Q}^T \cdot \mathbf{Q} \cdot \hat{\mathbf{T}}^3 \cdot \mathbf{Q}^T \cdot \mathbf{Q} \cdot \hat{\mathbf{T}}^3 \cdot \mathbf{Q}^T = \mathbf{T}^3$$

Therefore we have

$$\mathbf{T}^3 - I_T\mathbf{T}^2 + II_T\mathbf{T} - III_T\mathbf{I} = \mathbf{0}$$

Q.e.d

Example 2.5.3—Isotropic Tensor Function

Based on Cayley–Hamilton theorem any tensor-valued tensor function $\mathbf{F(A)}$ can be expressed as

$$\mathbf{F(A)} = \phi_0\mathbf{I} + \phi_1\mathbf{A} + \phi_2\mathbf{A}^2$$

With ϕ_0, ϕ_1, ϕ_2 are scalar valued function of invariants of the tensor $\mathbf{A}$. These functions are also called the integrity basis of the isotropic tensor functions.

If $\mathbf{U}$ is the right stretch tensor, then another strain measure describing the deformation, the natural logarithm of the right stretch tensor, sometimes is called the Hencky strain tensor, can be expressed as an isotropic tensor function

$$\ln(\mathbf{U}) = \phi_0\mathbf{I} + \phi_1\mathbf{U} + \phi_2\mathbf{U}^2$$

Determine the scalar value functions ϕ_0, ϕ_1, ϕ_2 for a complete description of the Hencky strain tensor.

Solution

The determination of the integrity basis ϕ_0, ϕ_1, ϕ_2 of the isotropic tensor function $\ln(\mathbf{U})$ can be done in the principle space, i.e. based on the principle directions or eigenvectors. Assuming that $U_I, U_I I and U_I II$ are the principle values of the right stretch tensor $\mathbf{U}$, then the associated Hencky strain tensor can be written as

$$\begin{pmatrix} \ln U_I & 0 & 0 \\ 0 & \ln U_{II} & 0 \\ 0 & 0 & \ln U_{III} \end{pmatrix} = \phi_0 \begin{pmatrix} 1 & 0 & 0 \\ 0 & 1 & 0 \\ 0 & 0 & 1 \end{pmatrix} + \phi_1 \begin{pmatrix} U_I & 0 & 0 \\ 0 & U_{II} & 0 \\ 0 & 0 & U_{III} \end{pmatrix}$$

$$+ \phi_2 \begin{pmatrix} U_I^2 & 0 & 0 \\ 0 & U_{II}^2 & 0 \\ 0 & 0 & U_{III}^2 \end{pmatrix}$$

Now there are three equations with the three unknown integrity bases.

$$\begin{pmatrix} 1 & U_I & U_I^2 \\ 1 & U_{II} & U_{II}^2 \\ 1 & U_{III} & U_{III}^2 \end{pmatrix} \begin{pmatrix} \phi_0 \\ \phi_1 \\ \phi_2 \end{pmatrix} = \begin{pmatrix} \ln U_I \\ \ln U_{II} \\ \ln U_{III} \end{pmatrix}$$

or

$$\begin{pmatrix} \phi_0 \\ \phi_1 \\ \phi_2 \end{pmatrix} = \begin{pmatrix} 1 & U_I & U_I^2 \\ 1 & U_{II} & U_{II}^2 \\ 1 & U_{III} & U_{III}^2 \end{pmatrix}^{-1} \begin{pmatrix} \ln U_I \\ \ln U_{II} \\ \ln U_{III} \end{pmatrix}$$

Problem 2.5.1

The following is a spectral form of a strain tensor $\hat{\mathbf{T}}$ given as

$$\hat{\mathbf{T}} = \begin{pmatrix} 2 & 0 & 0 \\ 0 & 1 & 0 \\ 0 & 0 & -3 \end{pmatrix}$$

Perform an orthogonal transformation with the following orthogonal tensor $\mathbf{Q}$

$$\mathbf{Q} = \frac{1}{3} \begin{pmatrix} 1 & 2 & 2 \\ 2 & 1 & -2 \\ -2 & 2 & -1 \end{pmatrix}$$

Show that these are two different forms of the same stain tensor, and hence show that the expression

$$\mathbf{T}^3 - II_T \mathbf{T}^2 + II_T \mathbf{T} - III_T \mathbf{I}$$

is a zero tensor.

Problem 2.5.2

Calculate the value of the following isotropic tensor function

$$\sin \begin{pmatrix} 1 & 3 & 2 \\ 3 & 1 & 5 \\ 2 & 5 & 2 \end{pmatrix}$$

2.6   Distortion and Dilatation

The deformation of a body or a volume element is composed of two parts, the one is distortion and the other one is dilatation. Distortion is deformation of a body without any change in volume but shape, while dilatation is associated with change in volume. These two types of deformation are represented by additive decomposition of strain tensor into deviatoric $\hat{\mathbf{E}}'$ and spherical or hydrostatic parts $\frac{1}{3} tr(\hat{\mathbf{E}})\mathbf{I}$. That is

$$\hat{\mathbf{E}} = \hat{\mathbf{E}}' + \frac{1}{3} tr(\hat{\mathbf{E}})\mathbf{I} \longleftrightarrow \epsilon_{ij} = \epsilon'_{ij} + \frac{1}{3}\epsilon_{kk}\delta_{ij}$$

The deviatoric part of strain tensor is responsible for the change of shape and the volume change is described by hydrostatic part of strain tensor.

2.6.1   Examples and Problems

Example 2.6.1

Given is the following St. Venant torsion

$$x_1 = X_1 - \frac{\theta}{l} X_2 X_3$$

$$x_2 = X_2 + \frac{\theta}{l} X_3 X_1$$

$$x_3 = X_3 + \psi(X_1, X_2)$$

Investigate the above motion if it deals with dilatation or distortion.

Solution

The displacement vector u_i can be obtained directly using the equation

$$u_i = x_i - X_i$$

which produces

$$u_1 = -\frac{\theta}{l}X_2X_3$$

$$u_2 = \frac{\theta}{l}X_3X_1$$

$$u_3 = \psi(X_1, X_2)$$

the components of the displacement vector. The corresponding displacement gradient **H**

$$(\mathbf{H}) = \partial u_i/\partial X_j = \begin{pmatrix} 0 & -(\theta/l)X_3 & -(\theta/l)X_2 \\ (\theta/l)X_3 & 0 & (\theta/l)X_1 \\ \partial\psi(X_1, X_2)/\partial X_1 & \partial\psi(X_1, X_2)/\partial X_2 & 0 \end{pmatrix}$$

and

$$\mathbf{H} + \mathbf{H}^{\mathsf{T}} = \begin{pmatrix} 0 & 0 & \partial\psi/\partial X_1 - (\theta/l)X_2 \\ 0 & 0 & \partial\psi/\partial X_2 + (\theta/l)X_1 \\ \partial\psi/\partial X_1 - (\theta/l)X_2 & \partial\psi/\partial X_2 + (\theta/l)X_1 & 0 \end{pmatrix}$$

Therefore the linearized Lagrangian strain tensor has the following form

$$(\mathbf{E}) = \frac{1}{2}\begin{pmatrix} 0 & 0 & \partial\psi/\partial X_1 - (\theta/l)X_2 \\ 0 & 0 & \partial\psi/\partial X_2 + (\theta/l)X_1 \\ \partial\psi/\partial X_1 - (\theta/l)X_2 & \partial\psi/\partial X_2 + (\theta/l)X_1 & 0 \end{pmatrix}$$

The hydrostatic part of linearized Lagrange strain tensor is 0 because

$$tr(\mathbf{E}) = 0$$

So in the St. Venant torsion there is no dilatation, i.e. no change in volume. Thus the deviatoric part $\hat{\mathbf{E}}'$ is **E** itself

$$\hat{\mathbf{E}}' = \frac{1}{2}\begin{pmatrix} 0 & 0 & \partial\psi/\partial X_1 - (\theta/l)X_2 \\ 0 & 0 & \partial\psi/\partial X_2 + (\theta/l)X_1 \\ \partial\psi/\partial X_1 - (\theta/l)X_2 & \partial\psi/\partial X_2 + (\theta/l)X_1 & 0 \end{pmatrix}$$

So the deformation of body during the St. Venant torsion is merely a change in shape of cross section due to shear and warping.

Problem 2.6.1

One particular point of a deformable body, the displacement gradient is measured and to have the following form

$$(\nabla \mathbf{u}) = \begin{pmatrix} 1 & -3 & 2 \\ 2 & 1 & 5 \\ 0 & -3 & 2 \end{pmatrix} \times 10^{-3}$$

1. Find the infinitesimal strain and spin tensor.
2. Calculate the spherical and deviatoric part of the infinitesimal stain tensor.
3. Calculate the associated eigenvalues and eigenvector of the infinitesimal strain tensor.

2.7 Deformation Velocity Tensor

Until now the discussion on kinematics of deformation was made without taking time into consideration. The strain tensor derived in previous section is very useful for describing the deformation of solid body but does not fit well in describing the deformation of fluid, liquid or gaseous material. For this particular purpose, we will use the concept of the deformation rate tensor or strain rate tensor. If any two particles in a continuous body move relatively to each other, then the velocity of one particle relative to neighbouring particle is understood as the of rate of change of velocity field with respect to the current position or current spatial coordinates. Mathematically, it is defined as follows

$$d\mathbf{v} = \frac{\partial \mathbf{v}}{\partial \mathbf{x}} \cdot d\mathbf{x} \tag{2.19}$$

or in short

$$d\mathbf{v} = \mathbf{L} \cdot d\mathbf{x}$$

The tensor $\mathbf{L} = \partial \mathbf{v}/\partial \mathbf{x} = \nabla \mathbf{v}$ is known as the velocity gradient tensor. This tensor is neither symmetry nor antimetry but can be splitted additively into symmetry and antimetry part by averaging it with its transpose $\mathbf{L}^{\mathrm{T}}$ or by subtracting out its transpose.

$$\mathbf{D} = 1/2(\mathbf{L} + \mathbf{L}^{\mathrm{T}}) = \mathbf{D}$$

$$\mathbf{W} = 1/2(\mathbf{L} - \mathbf{L}^{\mathrm{T}}) = -\mathbf{W}$$

The tensor $\mathbf{D}$ is known as rate of deformation tensor or also as stretching tensor while $\mathbf{W}$ is called vorticity or spin tensor. The deformation velocity tensor $\dot{\mathbf{F}}$ can be obtained by differentiating the deformation gradient tensor with respect to time. It can be shown that this tensor is indeed the velocity gradient tensor.

$$\dot{\mathbf{F}} = \frac{d\mathbf{F}}{dt}$$

with

$$\mathbf{F} = \frac{\partial \mathbf{x}}{\partial \mathbf{X}}$$

we have

$$\dot{\mathbf{F}} = \frac{d(\partial \mathbf{x}/\partial \mathbf{X})}{dt}$$

or

$$\dot{\mathbf{F}} = \frac{\partial \mathbf{v}}{\partial \mathbf{X}}$$

After applying the chain rule we have

$$\dot{\mathbf{F}} = \frac{\partial \mathbf{v}}{\partial \mathbf{x}} \cdot \frac{\partial \mathbf{x}}{\partial \mathbf{X}}$$

$$\dot{\mathbf{F}} = \nabla \mathbf{v} \cdot \mathbf{F} = \mathbf{L} \cdot \mathbf{F} \qquad (2.20)$$

Equation (2.20) describes the relationship between the deformation velocity tensor $\dot{\mathbf{F}}$ and velocity gradient tensor $\mathbf{L}$.

Velocity gradient tensor $\mathbf{L}$ describes the property of the velocity field of all particles in a body. It tells us how fast or slow the velocity vector varies with the current position. It is therefore Eulerian quantity. Figure 2.6 illustrates the concept of velocity gradient tensor. For the sake of simplicity, we take a case of stationary flow, i.e. $\partial \mathbf{v}/\partial t = 0$, the streamlines and pathlines are overlapping each other. The particles A and B move relative to each other as shown in Fig. 2.6 on two different stream lines. The relative velocity of particle A with regard to the particle B is $d\mathbf{v}$ and it is given by the multiplication of velocity slope or gradient and the distance between these two particles, i.e.

$$d\mathbf{v} = (\partial \mathbf{v}/\partial \mathbf{x}) \cdot d\mathbf{x}$$

This velocity slope or velocity gradient is called velocity gradient tensor $\mathbf{L} = \partial \mathbf{v}/\partial \mathbf{x}$.

In continuum mechanics, there are two different time derivatives describing the rate of change any field property with tensorial quantity of second order. The first one is called Jaumann time derivative and the other one is called the Oldroyd time derivative.

Let Φ be the property of the field, then the Jaumann time derivative of this property is defined as follows:

$$\frac{D_J \Phi}{Dt} = \frac{\partial \Phi}{\partial t} + \Phi \cdot \mathbf{W} - \mathbf{W} \cdot \Phi$$

and the Oldroyd time derivative is

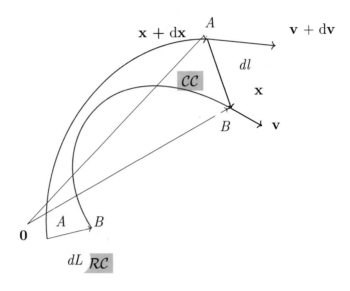

Fig. 2.6 Geometrical interpretation of velocity gradient **L**

$$\frac{D_O \Phi}{Dt} = \frac{\partial \Phi}{\partial t} + \Phi \cdot \mathbf{L} + \cdot \Phi^T \cdot \mathbf{L}$$

These two time derivatives are objective time derivatives. In other words, their invariants are the same for a stationary or for a moving observer.

2.7.1   Examples and Problems

Example 2.7.1

A motion of a continuum is given by the following relation

$$x_1 = X_1 e^t + X_3(e^t - 1), \; x_2 = X_2 + X_3(e^t - e^{-t}), \; x_3 = X^3$$

Determine the velocity field of this object in Lagrangian or material form. Calculate its velocity gradient tensor **L**.

Solution

The associated deformation gradient

$$(\mathbf{F}) = \begin{pmatrix} e^t & 0 & e^t - 1 \\ 0 & 1 & e^t - e^{-t} \\ 0 & 0 & 1 \end{pmatrix}$$

and the velocity field is

$$\dot{x}_1 = v_1 = X_1 e^t + X_3 e^t$$
$$\dot{x}_2 = v_2 = X_3(e^t + e^{-t})$$
$$\dot{x}_3 = v_3 = 0$$

The velocity gradient tensor $\mathbf{L}$ is calculated from the equation

$$\dot{\mathbf{F}} = \nabla \mathbf{v} \cdot \mathbf{F} = \mathbf{L} \cdot \mathbf{F}$$

The time derivative of the deformation gradient is

$$\dot{\mathbf{F}} = \begin{pmatrix} e^t & 0 & e^t \\ 0 & 0 & e^t + e^{-t} \\ 0 & 0 & 0 \end{pmatrix}$$

$$\mathbf{L} = \dot{\mathbf{F}} \cdot \mathbf{F}^{-1}$$

$$\mathbf{L} = \begin{pmatrix} e^t & 0 & e^t \\ 0 & 0 & e^t + e^{-t} \\ 0 & 0 & 0 \end{pmatrix} \begin{pmatrix} e^t & 0 & e^t - 1 \\ 0 & 1 & e^t - e^{-t} \\ 0 & 0 & 1 \end{pmatrix}^{-1}$$

Therefore the velocity gradient tensor is

$$\mathbf{L} = \begin{pmatrix} 1 & 0 & 1 \\ 0 & 0 & e^t - e^{-t} \\ 0 & 0 & 0 \end{pmatrix}$$

Problem 2.7.1

The motion of a deformable body is given by the following relations

$$x_1 = X_1$$
$$x_2 = (1/2)[(X_2 + X_3)\exp(\alpha)t] + (1/2)[(X_2 - X_3)\exp(-\alpha)t]$$
$$x_3 = (1/2)[(X_2 + X_3)\exp(\alpha)t] - (1/2)[(X_2 - X_3)\exp(-\alpha)t]$$

Determine the velocity field of this object in Lagrangian or material form. Calculate its velocity gradient tensor **L**.

Problem 2.7.2

The following relations describe the planar velocity field of a fluid motion.

$$v_1 = x_1$$
$$v_2 = 0$$
$$v_3 = \frac{x_3}{(1 + \kappa t)}$$

Determine the path and stream lines of the motion at the time t_0 and passing through the point P with the coordinates (P_1, P_2, P_3).

2.8 Condition of Compatibility

From a displacement field, one can with ease find the Lagrangian or Eulerian strain tensor, that is, by simply differentiating it with respect to material coordinates or with respect to spatial coordinates. However from known strain tensor the task of finding the associated displacement field is quite complex, that is a task of integrating the strain field with some resulting unknown integration functions and constants. Furthermore, the solution obtained from the integration is not necessarily unique as the displacement field, in contrast to strain measure, contains rigid body motion. There is then a need to derive auxiliary equations to determine those integration functions and constants to fit the displacement field completely and uniquely. These auxiliary conditions are called the compatibility or integrability conditions. The physical interpretation of condition of compatibility is what are the conditions required to bring two independent set of particles in a continuum displaced before and after the deformation, so that they fit together seamlessly. Figure 2.7 illustrates the concept of compatibility conditions or compatible strains. The compatibility conditions for the strain field are obtained by eliminating the displacement functions from the definition equation for the Lagrangian strain tensor, i.e. from Eq. (2.16) or the indicial version of it.

$$\mathbf{E} = \frac{1}{2}(\mathbf{H} + \mathbf{H}^{\mathrm{T}} + \mathbf{H}^{\mathrm{T}} \cdot \mathbf{H})$$

$$\epsilon_{ij} = \frac{1}{2}\left(\frac{\partial u_i}{\partial X_j} + \frac{\partial u_j}{\partial X_i} + \frac{\partial u_j}{\partial X_k}\frac{\partial u_i}{\partial X_k}\right)$$

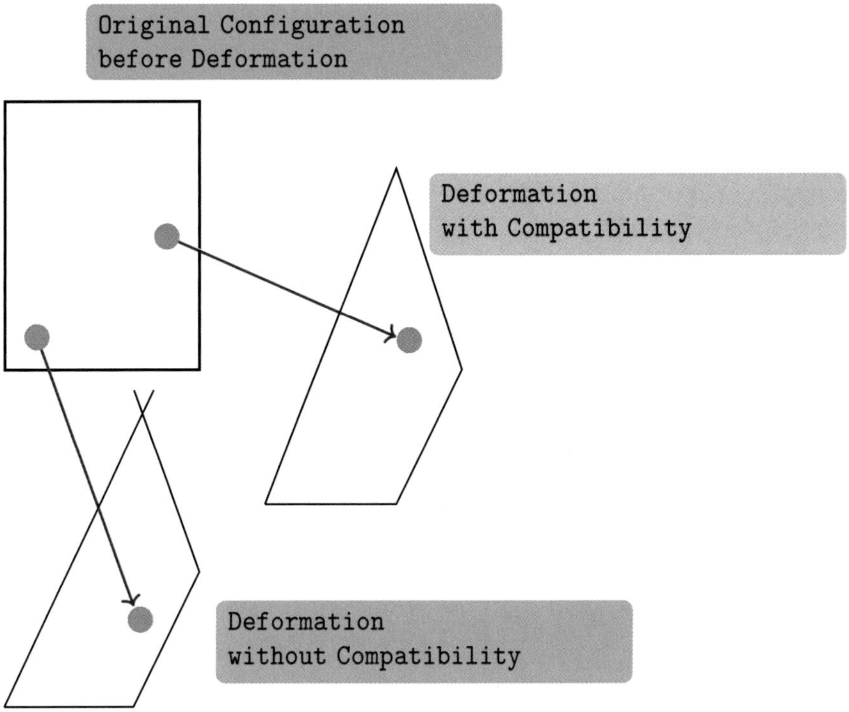

Original Configuration
before Deformation

Deformation
with Compatibility

Deformation
without Compatibility

Fig. 2.7 Geometrical Interpretation of Compatibility Conditions

For small strains, there is no need to differentiate between the usage of material coordinates X_i and spatial coordinate x_j, so that one can start from the classical infinitesimal strain tensor

$$\hat{\mathbf{E}} = \frac{1}{2}(\mathbf{H} + \mathbf{H}^{\mathrm{T}})$$

$$\epsilon_{ij} = \frac{1}{2}\left(\frac{\partial u_i}{\partial X_j} + \frac{\partial u_j}{\partial X_i}\right) = \frac{1}{2}\left(\frac{\partial u_i}{\partial x_j} + \frac{\partial u_j}{\partial x_i}\right) = \frac{1}{2}(u_{i,j} + u_{j,i})$$

By differentiation and exchanging the indices, one obtains

$$\epsilon_{ij,kl} + \epsilon_{kl,ij} - \epsilon_{ik,jl} - \epsilon_{jl,ik} = 0 \tag{2.21}$$

After expansion of the above Eq. (2.21) using the conventional Eulerian spatial coordinates $x_1 = x$, $x_2 = y$ and $x_3 = z$ the compatibility conditions for the three-dimensional case take the following form

$$\frac{\partial^2 \epsilon_{yy}}{\partial^2 z} + \frac{\partial^2 \epsilon_{zz}}{\partial^2 y} = 2\frac{\partial^2 \epsilon_{yz}}{\partial z \partial y}$$

$$\frac{\partial^2 \epsilon_{zz}}{\partial^2 x} + \frac{\partial^2 \epsilon_{xx}}{\partial^2 z} = 2\frac{\partial^2 \epsilon_{xz}}{\partial z \partial x}$$

$$\frac{\partial^2 \epsilon_{xx}}{\partial^2 y} + \frac{\partial^2 \epsilon_{yy}}{\partial^2 x} = 2\frac{\partial^2 \epsilon_{xy}}{\partial x \partial y}$$

$$\frac{\partial^2 \epsilon_{zz}}{\partial x \partial y} = \frac{\partial}{\partial z}\left(-\frac{\partial \epsilon_{xy}}{\partial z} + \frac{\partial \epsilon_{xz}}{\partial y} + \frac{\partial \epsilon_{yz}}{\partial x}\right) \qquad (2.22)$$

$$\frac{\partial^2 \epsilon_{yy}}{\partial x \partial z} = \frac{\partial}{\partial y}\left(-\frac{\partial \epsilon_{xz}}{\partial y} + \frac{\partial \epsilon_{yz}}{\partial x} + \frac{\partial \epsilon_{xy}}{\partial z}\right)$$

$$\frac{\partial^2 \epsilon_{xx}}{\partial z \partial y} = \frac{\partial}{\partial x}\left(-\frac{\partial \epsilon_{yz}}{\partial x} + \frac{\partial \epsilon_{xz}}{\partial y} + \frac{\partial \epsilon_{yz}}{\partial x}\right)$$

The above six compatibility conditions (Eq. 2.22) are the necessary conditions for the components of the strain tensor to be continuum mechanically compatible, i.e. to be considered as measure for deformation (refer Fig. 2.7). For the two-dimensional case, the components of strain tensor must fulfil the following modified compatibility condition.

$$\frac{\partial^2 \epsilon_{xx}}{\partial^2 y} + \frac{\partial^2 \epsilon_{yy}}{\partial^2 x} = 2\frac{\partial^2 \epsilon_{xy}}{\partial x \partial y} \qquad (2.23)$$

2.8.1   Examples and Problems

Example 2.8.1

Given is the following two-dimensional displacement field of a body.

$$u_1 = Ax_1 x_2$$

$$u_2 = Bx_2 x_2$$

where A and B are small constants. Determine

1. the components of linearized (small) strain,
2. whether the strains obey the compatibility condition.

Solution

For small strains, one starts from

$$\epsilon_{ij} = \frac{1}{2}\left(\frac{\partial u_i}{\partial x_j} + \frac{\partial u_j}{\partial x_i}\right) = \frac{1}{2}(u_{i,j} + u_{j,i})$$

and get the linearized strain tensor

$$(\mathbf{E}) = \epsilon_{ij} = \frac{1}{2}\begin{pmatrix} 2Ax_2 & Ax_1 \\ Ax_1 & 4Bx_2 \end{pmatrix}$$

With the above tensor of small strain the compatibility condition for two-dimensional case, Eq. (2.23) is automatically fulfilled.

Problem 2.8.1

A square plate is clamped support along x- and y-axis and loaded with externally distributed load along its diagonal. From the measurement data taken by strain gauges, it is assumed that the components of strain tensor take the following form

$$\epsilon_{xx} = Ax^2y + y^3$$

and

$$\epsilon_{yy} = Bxy^2$$

Based on this assumption determine the displacement field and the slip γ_{xy}. What is the condition for the fulfilment of the compatibility conditions?

2.9 Deformation Gradient of a Discretized Body

The basic idea of the finite element method (FEM) is to discretize continuum into a accumulation of well-defined elements. Thus the determination of the displacement field and the deformation of the continuum is reduced to the establishing the motion or the deformation of each element of discretized body. All the equations related to the state of strain, state of stress and constitutive law of a body are constructed based on the defined element and solving these equations delivers the information about the deformation of the element. The total deformation of the discretized body is then the integration of the individual element deformation through the application of the corresponding compatibility equations.

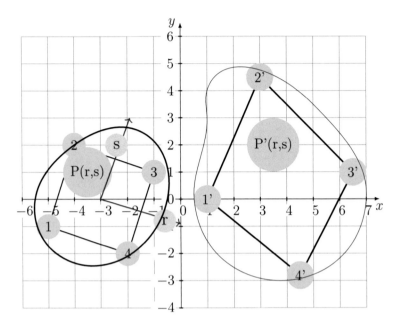

Fig. 2.8 Discretized continuum: before and after deformation

The weighted combination of the motion of the nodes defines the movement of any point within the element. This in return provides the sought displacement field of the element, i.e. the displacement of any point within the considered element.

Figure 2.8 illustrates the deformation of the discretized two-dimensional continuum, represented by a plane Ω bounded by a closed curve $\partial\Omega$. The plane is approximated by a single element, a quadlateral with 4 vertices or nodes $i, i = 1, 2, 3, 4$ After deformation the plane Ω takes a new configuration represented by quadrilateral with nodes $i', i = 1, 2, 3, 4$. Let $P(r, s)$ be the position of any material point P in the reference configuration, i.e. undeformed configuration described using the local cartesian coordinate system $\xi = (r, s)$. In term of global cartesian coordinate system, the position $P(x, y)$, $\mathbf{x} = (x, y)$ can be related by the following transformation equations

$$x(r, s) = \phi_1(r, s)x_1 + \phi_2(r, s)x_2 + \phi_3(r, s)x_3 + \phi_4(r, s)x_4 \qquad (2.24)$$
$$y(r, s) = \phi_1(r, s)y_1 + \phi_2(r, s)y_2 + \phi_3(r, s)y_3 + \phi_4(r, s)y_4 \qquad (2.25)$$

The position of the point P is then the linear combination of the position of nodes $i, i = 1, 2, 3, 4$ linearly interpolated by the weight or "shape" functions $\phi_i(r, s)$. For two-dimensional case the shape functions are defined as follows

$$\phi_1(r, s) = \frac{1}{4}(1 - r)(1 - s)$$

$$\phi_2(r, s) = \frac{1}{4}(1 + r)(1 - s)$$

$$\phi_3(r, s) = \frac{1}{4}(1 + r)(1 + s)$$

$$\phi_4(r, s) = \frac{1}{4}(1 - r)(1 + s)$$

The same is the case for the displacement of any point within Ω. In terms of the displacement of the nodes i, $i = 1, 2, 3, 4$ the displacement of the material point $P(r, s)$ is given as follows:

$$u(r, s) = \phi_1(r, s)u_1 + \phi_2(r, s)u_2 + \phi_3(r, s)u_3 + \phi_4(r, s)u_4 \qquad (2.26)$$
$$v(r, s) = \phi_1(r, s)v_1 + \phi_2(r, s)v_2 + \phi_3(r, s)v_3 + \phi_4(r, s)v_4 \qquad (2.27)$$

Above equations describe the displacement field of the discretized continuum can be determined and the deformation gradient can be obtained from these equations.

The global deformation gradient, $\mathbf{F}$, i.e. the deformation gradient expressed in the global coordinates, can be obtained by determining the local displacement gradient $\partial \mathbf{u}/\partial \xi$, with $\mathbf{u} = (u, v)$ and $\xi = (r, s)$. By applying the chain rule the global displacement gradient can be given as

$$\partial \mathbf{u}/\partial \mathbf{x} = (\partial \mathbf{u}/\partial \xi)(\partial \xi/\partial \mathbf{x})$$

The global deformation gradient $\mathbf{F}$ can be calculated from

$$\mathbf{F} = \mathbf{I} + \partial \mathbf{u}/\partial \mathbf{x}$$

For the case of three-dimensional continuum, the discretization of body is done by selecting linear or curvilinear tetrahedral as primary element, depending the complexity of the body to be discretized.

2.9.1 Examples and Problems

Example 2.9.1

Figure 2.8 shows a bounded plane is approximated by a quadlateral with vertices $1 : (-5, -1)$, $2 : (-4, 2)$, $3 : (-1, 1)$ and $4 : (-2, -2)$. After deformation the vertices displace to the position $1' : (1, 0)$, $2' : (3, 4)$, $3' : (6.5, 1)$ and $4' : (4.5, -2.5)$.

Calculate the displacement field of the plane and compute the displacement vector of a point $P(-3.5, 1)$. Express the position of the point P in global coordinates x, y. Calculate the global deformation gradient $\mathbf{F}$ and identify the rotation and the stretching part in the deformation.

Solution

The position of the vertices in the reference configuration is listed below

$$x_1 = -5, x_2 = -4, x_3 = -1, x_4 = -2$$

$$y_1 = -1, y_2 = 2, y_3 = 1, y_4 = -2$$

The displacement vector of the vertices or nodes are

$$u_1 = 6, u_2 = 7, u_3 = 7.5, u_4 = 6.5$$

$$v_1 = 1, v_2 = 2, v_3 = 0, v_4 = -0.5$$

In term of global coordinates (x, y) the point $P(-3.5, 1)$ can be expressed as

$$x = 1/4(1-r)(1-s)x1 + 1/4(1+r)(1-s)x2 + 1/4(1+r)(1+s)x3$$
$$+ 1/4(1-r)(1+s)x4$$

$$y = 1/4(1-r)(1-s)y1 + 1/4(1+r)(1-s)y2$$
$$+ 1/4(1+r)(1+s)y3 + 1/4(1-r)(1+s)y4$$

Upon substitution produces

$$x_P = -0.17125, \quad y_P = 1.1$$

The displacement field in terms of local coordinates (r, s) is given as follows:

$$u(r, s) = \frac{1}{4}(1-r)(1-s)u_1 + \frac{1}{4}(1+r)(1-s)u_2 + \frac{1}{4}(1+r)(1+s)u_3 + \frac{1}{4}(1-r)(1+s)u_4$$

and

$$v(r, s) = \frac{1}{4}(1-r)(1-s)v_1 + \frac{1}{4}(1+r)(1-s)v_2$$
$$+ \frac{1}{4}(1+r)(1+s)v_3 + \frac{1}{4}(1-r)(1+s)v_4$$

Substituting the individual displacement of the nodes, we have the displacement field of the element

$$u(r, s) = \frac{3}{2}(1 - r)(1 - s) + \frac{7}{4}(1 + r)(1 - s) + \frac{15}{8}(1 + r)(1 + s) + \frac{13}{8}(1 - r)(1 + s)$$

$$v(r, s) = \frac{1}{4}(1 - r)(1 - s) + \frac{1}{2}(1 + r)(1 - s) - \frac{1}{8}(1 - r)(1 + s)$$

For the displacement of the point $P(-3.5, 1)$ we set $r = -3.5$ and $s = 1$, we obtain

$$u(-3.5, 1) = \frac{3}{2}(1 - (-3.5))(1 - (1)) + \frac{7}{4}(1 + (-3.5))(1 - (1)) + \frac{15}{8}(1 + (-3.5))(1 + (1))$$

$$+ \frac{13}{8}(1 - (-3.5))(1 + (1))$$

$$= 5.875$$

$$v(r, s) = \frac{1}{4}(1 - (-3.5))(1 - (1)) + \frac{1}{2}(1 - (-3.5))(1 - (1)) - \frac{1}{8}(1 - (-3.5))(1 - (1))$$

$$= -1.125$$

Therefore, the displacement of the material point P to P' is

$$\mathbf{u}_P = (5.875, -1.125)$$

From the displacement field obtained previously, the elements of the displacement gradient in local coordinate are calculated as follows:

$$\partial u / \partial r = -\frac{1}{4}(1 - s)u_1 + \frac{1}{4}(1 - s)u_2 + \frac{1}{4}(1 + s)u_3 - \frac{1}{4}(1 + s)u_4$$

$$\partial u / \partial s = -\frac{1}{4}(1 - r)u_1 - \frac{1}{4}(1 + r)u_2 + \frac{1}{4}(1 + r)u_3 + \frac{1}{4}(1 - r)u_4$$

$$\partial v / \partial r = -\frac{1}{4}(1 - s)v_1 + \frac{1}{4}(1 - s)v_2 + \frac{1}{4}(1 + s)v_3 - \frac{1}{4}(1 + s)v_4$$

$$\partial v / \partial s = -\frac{1}{4}(1 - r)v_1 + \frac{1}{4}(1 - r)v_2 + \frac{1}{4}(1 + r)v_3 - \frac{1}{4}(1 + r)v_4$$

One can obtain the global displacement gradient performing the coordinate transform between these two coordinate systems. We get first the global and local coordinates transformation

$$\partial x/\partial r = -\frac{1}{4}(1-s)x_1 + \frac{1}{4}(1-s)x_2 + \frac{1}{4}(1+s)x_3 - \frac{1}{4}(1+s)x_4$$

$$\partial x/\partial s = -\frac{1}{4}(1-r)x_1 - \frac{1}{4}(1+r)x_2 + \frac{1}{4}(1+r)x_3 + \frac{1}{4}(1-r)x_4$$

$$\partial y/\partial r = -\frac{1}{4}(1-s)y_1 + \frac{1}{4}(1-s)y_2 + \frac{1}{4}(1+s)y_3 - \frac{1}{4}(1+s)y_4$$

$$\partial y/\partial s = -\frac{1}{4}(1-r)y_1 + \frac{1}{4}(1-r)y_2 + \frac{1}{4}(1+r)y_3 - \frac{1}{4}(1+r)y_4$$

Now after applying the chain rule the global displacement gradient $\partial\mathbf{u}/\partial\mathbf{x}$ is given as

$$\partial\mathbf{u}/\partial\mathbf{x} = \partial\mathbf{u}/\partial\xi \cdot \partial\xi/\partial\mathbf{x} = \partial\mathbf{u}/\partial\xi \cdot (\partial\mathbf{x}/\partial\xi)^{-1}$$

or in matrix form

$$\begin{pmatrix} \partial u/\partial x & \partial u/\partial y \\ \partial v/\partial x & \partial v/\partial y \end{pmatrix} = \begin{pmatrix} \partial u/\partial r & \partial u/\partial s \\ \partial v/\partial r & \partial v/\partial s \end{pmatrix} \begin{pmatrix} \partial r/\partial x & \partial r/\partial y \\ \partial s/\partial x & \partial s/\partial y| \end{pmatrix}$$

$$= \begin{pmatrix} \partial u/\partial r & \partial u/\partial s \\ \partial v/\partial r & \partial v/\partial s \end{pmatrix} \begin{pmatrix} \partial x/\partial r & \partial x/\partial s \\ \partial y/\partial r & \partial y/\partial s \end{pmatrix}^{-1}$$

By substituting numerical values used above the matrix equation can be written as

$$\begin{pmatrix} \partial u/\partial x & \partial u/\partial y \\ \partial v/\partial x & \partial v/\partial y \end{pmatrix} = \begin{pmatrix} 0.5 & 0.25 \\ 0.25 & -3.94 \end{pmatrix} \begin{pmatrix} 6. & 4. \\ -1.33 & -1.33 \end{pmatrix}$$

$$\partial\mathbf{u}/\partial\mathbf{x} = \begin{pmatrix} 2.67 & 1.67 \\ 6.75 & 6.25 \end{pmatrix}$$

The global deformation gradient at the point $P(-3.5, 1)$ is then calculated from

$$\mathbf{F} = \mathbf{I} + \partial\mathbf{u}/\partial\mathbf{x}$$

$$\mathbf{F} = \begin{pmatrix} 3.67 & 1.67 \\ 6.75 & 7.25 \end{pmatrix}$$

Rotation or stretching of the element can be obtained from the polar decomposition of the deformation gradient,

$$\mathbf{F} = \mathbf{R} \cdot \mathbf{U} = \mathbf{V} \cdot \mathbf{R}$$

The left or right stretch tensors, **V** or **U** can be determined from

$$\mathbf{V}^2 = \mathbf{F} \cdot \mathbf{F}^T$$

or

$$\mathbf{U}^2 = \mathbf{F}^T \cdot \mathbf{F}$$

With the above-calculated deformation gradient, both stretch tensors are

$$\mathbf{U}^2 = \begin{pmatrix} 52.72 & 46.66 \\ 46.66 & 41.88 \end{pmatrix}$$

and 52.7186, 46.6753, 46.6753, 41.882

$$\mathbf{V}^2 = \begin{pmatrix} 9.89 & 28.43 \\ 28.43 & 84.71 \end{pmatrix}$$

As we can see, the right and left stretch tensors and their squares are not equal and the difference between these two stretch tensors is significantly small. This suggests that there is a rotation of the element in the deformation process besides the stretching. However, the rotation is relatively minimal. We leave it to the reader to find the associated rotation tensor **R** in this deformation.

Problem 2.9.1

A quadrilateral originally with vertices at the points $1 : (2, 1), 2 : (-3, 3), 3 : (-1, 4)$ and $4 : (-2, -2)$. After deformation the vertices are displaced to the position $1' : (4, 5), 2' : (6, 4), 3' : (5, -3)$ and $4' : (4, -2)$. Calculate the displacement field of the plane and express the position of the point $P(1, 1)$ of an undeformed quadrilateral in term of the global coordinates x, y. Calculate the global deformation gradient **F** and the associated Lagrangian strain tensor **E**.

Chapter 3
Stress

3.1 Force and Stress

The concept of force is rather complex but C. Truesdell and W. Noll, amongst
co-founders of modern continuum mechanics, summarized it in "Encyclopedia of
Physics - Non linear Field Theories in Mechanics" as an action of the outside world
on a body in motion and the interaction between different parts of the body. In continuum mechanics, a body in motion is assumed to be acted upon by two type of forces,
body forces per unit mass of the body **b** and contact forces or surface forces **t** per unit
oriented area of contact surface $d\mathbf{A}$. Figure 3.1 illustrates all these vectors. Typical
examples of body forces are gravitational force, centrifugal forces and electrostatic
and magnetic forces while common example of surface or contact force is the surface
tension in the meniscus of mercury in a capillary tube.

The limit of the distributed contact force over unit oriented area of contact surface
as the area decreases to zero is called stress. Since forces are described as vectors,
so is the stress, σ, as shown in Fig. 3.1.

So if force is an action of the outside world on the body, then stress is the expression
of how this outside action of force is distributed in the body. The body therefore
experiences the state of stress which delivers stress vector in any direction. Since
in three-dimensional euclidean space there are three linearly independent directions
characterized by the outside normal vector to the associated contact surface, there are
then three stress vectors associated with these three linearly independent directions.

For the sake of simplicity, we use the conventional Cartesian basis vectors to
represent the direction of stress vectors in a body. By doing so we have σ_i, $i = 1, 2, 3$
for stress vector in the directions $\mathbf{x_1}$, $\mathbf{x_2}$ and $\mathbf{x_3}$, respectively.[1]

---

[1] For the reference configuration one uses the direction vectors of material coordinates $\mathbf{X_1}$, $\mathbf{X_2}$ and
$\mathbf{X_3}$ to obtain the corresponding stress vectors.

© The Author(s), under exclusive license to Springer Nature Singapore Pte Ltd. 2023
N. A. N. Mohamed, *Introduction to Continuum Mechanics for Engineers*,
https://doi.org/10.1007/978-981-99-0811-0_3

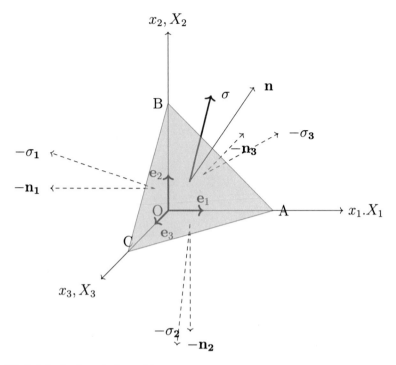

Fig. 3.1 Infinitesimal tetrahedron with stress vectors

3.2 Cauchy Stress Tensor

As mentioned earlier the stress is defined as a concentration of contact force over an oriented contact surface as the area of this surface mathematical decreases to zero. A question arises as the area of contact force is concerned. Since the body undergoes steadily deformation process, the contact area subsequently deforms as well. There is a need then to distinguish clearly the difference between stress based on initial contact area and the stress defined based on current oriented contact area. In material testing, common definition of stress is the limit of applied force over the sample initial cross-sectional area. This is well-accepted definition, and it is called an engineering stress or sometimes a nominal stress. True stress is called for the stress which is defined based on the current, undergoing deformation contact area. In continuum mechanics this stress is also known as Cauchy stress, while the other one is called Piola–Kirchhoff stress. Referring back to Fig. 3.1 and assuming that the tetrahedron is part of body under deformation, all the stresses to be derived are Cauchy stresses. Consider the front surface of the tetrahedron, represented by triangle ABC with infinitesimal area of dA. The surface is oriented by the normal vector $\mathbf{n}$ perpendicular to the surface, with $d\mathbf{A} = dA\mathbf{n}$ as oriented infinitesimal area. The stress associated with this surface is denoted by σ. The force associated with this stress is $\mathbf{t} = \sigma dA$ For the other surfaces of tetrahedron, triangles OBC, OAB and OAC the stress is denoted by σ_1, σ_2 and σ_3 respectively, so that the associated surface forces are $\mathbf{t}_1 = \sigma_1 dAn_1$ with dAn_1 is the projection of dA on $x_2 - x_3$ plane. Similarly goes for $\mathbf{t}_2 = \sigma_2 dAn_2$ and $\mathbf{t}_3 = \sigma_3 dAn_3$. With body force $\mathbf{b} = \rho dV$ the equation of motion is given as

$$\mathbf{t} - \mathbf{t}_1 - \mathbf{t}_2 - \mathbf{t}_3 + \mathbf{b}\rho dV = \mathbf{a}\rho dV$$

or in indicial form

$$t_i = t_{1i} - t_{2i} - t_{3i} + b_i \rho dV = a_i \rho dV$$

Expressed in terms of stresses

$$\sigma_i dA - \sigma_{1i} n_1 dA - \sigma_{2i} n_2 dA - \sigma_{3i} n_3 dA + + b_i \rho dV = a_i \rho dV$$

As the volume of the infinitesimal tetrahedron decreases to zero $dV \to 0$, we have from the above equation the following relation

$$\sigma_i dA = \sigma_{1i} n_1 dA + \sigma_{2i} n_2 dA + \sigma_{3i} n_3 dA$$

or

$$\sigma_i = \sigma_{1i} n_1 + \sigma_{2i} n_2 + \sigma_{3i} n_3$$

For free indices $i, j = 1, 2, 3$, the equation above can be summarized as

$$\sigma_i = \sigma_{ji} n_j \tag{3.1}$$

Symbolically, it takes the following form

$$\sigma = \mathbf{S}^T \cdot \mathbf{n} \tag{3.2}$$

Here is $\mathbf{S}^T$ the transpose of Cauchy stress tensor and σ is the stress vector associated with the oriented area characterized by the normal vector $\mathbf{n}$. In classical continuum mechanics, the symmetry of Cauchy stress tensor is secured and postulated in Boltzmann axiom, which states that the shear stress is pairwise assigned, i.e. $\sigma_{ij} = \sigma_{ji}$ for $i, j = 1, 2, 3$ or $\mathbf{S} = \mathbf{S}^T$. Therefore Eq. (3.2) can be also written as

$$\sigma = \mathbf{S} \cdot \mathbf{n}$$

So the above equation or Eq. (3.2) gives us the relationship between stress vector and stress tensor associated with specific orientation of contact area.

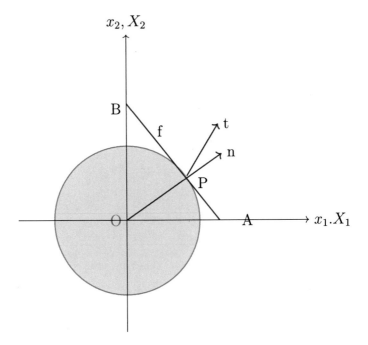

Fig. 3.2 Problem 3.1.2

3.2.1 *Examples and Problems*

Example 3.2.1

A cylindrical rod represented by a cylinder $x_1 x_1 + x_2 x_2 = 1$ is loaded with torsion moment so that the resulted shear stress takes the form $\sigma_{12} = 2\sigma$, where σ is a constant (refer to Fig. 3.2). The atmospheric pressure is assumed to be $\kappa\sigma$. Calculate traction vector **t** with respect to the tangential plane to the cylindrical rod at the point P(4/5, 3/5, 1). Calculate also the component vectors of **t** on that plane.

Solution

Due to the loading and atmospheric pressure, the state of stress in the rod represented by the Cauchy stress tensor **S** superposed by the hydrostatic (atmospheric) pressure and the torsion is

$$(\mathbf{S}) = \sigma_{ij} = \kappa\sigma \begin{pmatrix} 1 & 0 & 0 \\ 0 & 1 & 0 \\ 0 & 0 & 1 \end{pmatrix} + \begin{pmatrix} 0 & 2\sigma & 0 \\ 2\sigma & 0 & 0 \\ 0 & 0 & 0 \end{pmatrix}$$

$$(\mathbf{S}) = \sigma_{ij} = \kappa\sigma \begin{pmatrix} 1 & 2/\kappa & 0 \\ 2/\kappa & 1 & 0 \\ 0 & 0 & 1 \end{pmatrix}$$

The gradient of the cylindrical surface of the rod at any point is obtained from

$$\mathbf{f} = \nabla(x_1 x_1 + x_2 x_2 = 1) = (2x_1, 2x_2, 0)$$

For the point P(4/5,3/5,1), the gradient or slope vector is then

$$\mathbf{f} = (8/5, 6/5, 0)$$

a vector perpendicular to this gradient vector is then

$$\mathbf{g} = (6, -8, 0)$$

The normalized $\mathbf{g}$ is the normal vector to the above gradient vector, here

$$\mathbf{n} = 1/(100)(6, -8, 0)$$

Now we are to calculate the traction vector associated with the above normal vector, that is

$$\mathbf{t}/\kappa\sigma = \begin{pmatrix} & & & 6/100 \\ & & & -8/100 \\ & & & 0 \\ 1 & 2/\kappa & 0 & \frac{1}{100}(6-16/\kappa) \\ 2/\kappa & 1 & 0 & \frac{1}{100}(12-8/\kappa) \\ 0 & 0 & 1 & 0 \end{pmatrix}$$

which produces

$$\mathbf{t} = \frac{\kappa\sigma}{100}[(6-16/\kappa), (12/\kappa - 8), 0]$$

the sought traction vector. The component of the traction vector that is parallel to the normal vector is

$$\mathbf{t_n} = (\mathbf{t} \cdot \mathbf{n})\mathbf{n}$$

Therefore the component of $\mathbf{t}$, that is on the plane, i.e. tangential components is

$$\mathbf{t_t} = \mathbf{t} - (\mathbf{t} \cdot \mathbf{n})\mathbf{n}$$

or by introducing a dyadic product $\otimes$,[2] we have

$$\mathbf{t_t} = (\mathbf{I} - (\mathbf{n} \otimes \mathbf{n})) \cdot \mathbf{t}$$

---

[2] Refer to Appendix I.

Problem 3.2.1

A homogeneous heavy piece of wood moves uniformly along a horizontal plane with rough surface and the coefficient of dry friction $\mu_o = 0.35$. The area of the contact surface equals 0.012 m^2, and the mass of the wood block equals 10 kg. Find the stress vector on the plane under the wooden block.

3.3 Principal Values and Directions

We know by now that the stress tensor produces any traction or stress vector, $\mathbf{t}$ or σ, provided the normal vector of the oriented surface is known, as described by the Eq. (3.2) or

$$\sigma = \mathbf{S} \cdot \mathbf{n}$$

Now one question arises: Out of all oriented directions of the contact surfaces is there any specific direction of the surface that produces the traction or stress vector a multiple of the normal vector of that particular surface, i.e.

$$\sigma = \mathbf{S} \cdot \mathbf{n} = \lambda \mathbf{n} \tag{3.3}$$

where λ is a constant representing a simple multiplying factor. The above equation is basically asking for a special direction and multiplication of this normal vector that represent the whole state of stress in a loaded body. In continuum mechanics, we call these directions and multiplying factors as the principal direction and principal values. They are the solutions of the eigenvalue problem. The eigenvalue problem can be obtained directly from Eq. (3.3),

$$\det(\mathbf{S} - \lambda \mathbf{I}) = 0 \tag{3.4}$$

The expansion of the above equation produces the characteristic equation of the Cauchy stress tensor.

$$-\lambda^3 + I_1\lambda^2 - I_2\lambda + I_3 = 0 \tag{3.5}$$

The coefficients I_1, I_2 and I_3 are called the invariants of the stress tensor $\mathbf{S}$ and are given by

$$I_1 = tr\,(\mathbf{S}) = \sigma_{kk}$$

$$I_2 = \frac{1}{2}((tr\,\mathbf{S})^2 - tr\,(\mathbf{S} \cdot \mathbf{S})) = \frac{1}{2}(\sigma_{jj}\sigma_{kk} - \sigma_{jk}\sigma_{kj})$$

$$I_3 = \det(\mathbf{S})$$

The principal values λ are the solutions of the above characteristic polynomial for $\mathbf{S}$, whilst the principal directions are the eigenvectors calculated from Eq. (3.3) associated with the eigenvalues λ obtained previously.

3.3.1 Examples and Problems

Example 3.3.1

Given is the following Cauchy stress tensor

$$(\mathbf{S}) = \sigma_{ij} = \begin{pmatrix} 1 & 2 & 1 \\ 2 & 3 & 0 \\ 1 & 0 & -2 \end{pmatrix} \text{ MPa}$$

Calculate all the invariants and hence find the principal values of the tensor $\mathbf{S}$.

Solution

The first invariant is the trace of the tensor $\mathbf{S}$.

$$I_1 = tr\,(\mathbf{S}) = \sigma_{kk} = \sigma_{11} + \sigma_{22} + \sigma_{33} = 1 + 3 - 2 = 2$$

The second and third invariants can be calculated accordingly

$$I_2 = \frac{1}{2}((tr\,\mathbf{S})^2 - tr\,(\mathbf{S} \cdot \mathbf{S})) = \frac{1}{2}(\sigma_{jj}\sigma_{kk} - \sigma_{jk}\sigma_{kj})$$

$$I_3 = \det(\mathbf{S}) = -6 + 8 - 3 = -1$$

By utilizing matrix multiplication, we obtain

$$\mathbf{S} \cdot \mathbf{S} = \begin{pmatrix} 6 & 8 & -1 \\ 8 & 13 & 2 \\ -1 & 2 & 5 \end{pmatrix} (\text{MPa})^2$$

So the trace of $\mathbf{S} \cdot \mathbf{S}$ is

$$tr\,(\mathbf{S} \cdot \mathbf{S}) = 6 + 13 + 5 = 24$$

Therefore, the second invariant I_2 is

$$I_2 = \frac{1}{2}((tr\,\mathbf{S})^2 - tr\,(\mathbf{S} \cdot \mathbf{S}))$$

$$I_2 = \frac{1}{2}[(2)^2 - (24)] = -10$$

Now we substitute these invariants in the characteristic polynomial and obtain

$$-\lambda^3 + 2\lambda^2 + 10\lambda - 1 = 0$$

The principal values are obtained after solving the above polynomial and they are,

$$\lambda_1 = 4.98 \, \text{MPa}, \lambda_2 = 0.1 \, \text{MPa}, \lambda_3 = -2.38 \, \text{MPa}$$

An alternative solution is to find the determinant expressed in Eq. (3.4). Here we have

$$\det \begin{pmatrix} 1 - \lambda & 2 & 1 \\ 2 & 3 - \lambda & 0 \\ 1 & 0 & -2 - \lambda \end{pmatrix} = 0$$

After expansion and rearranging the component, we obtain

$$-\lambda^3 + 2\lambda^2 + 10\lambda - 1 = 0$$

the characteristic equation for the tensor **S**. Notice that the invariants are the coefficient of λ^0, λ^1 and λ^2 respectively.

Example 3.3.2

A pressure vessel shown in Fig. 3.3 is made of cylinder welded by caps at both ends. The cap is made of a deep drawn circular plate while the cylinder is made of rectangular plate rolled and welded at the end at the angle of 45° from horizontal axis. The inner diameter of cylinder is $D = 2R$ and the thickness of the plate is t. The vessel is filled with liquid of pressure p. Assuming that the vessel is a thin-walled vessel, i.e. $D/t > 10$, calculate the stresses in the vessel. Use your results to calculate the stress along the indicated weld beads or line (red line).

Numerical values: $p = 3 \, \text{MPa}$, $D = 2 \, \text{m}$, $L = 5 \, \text{m}$ and $t = 10 \, \text{mm}$

Solution

With the assumption that the vessel is thin-walled and the loading from gas pressure is isotropic, the stresses in the vessel resulted from the loading is only in axial and circumferential directions. Stress in radial direction is considered to be negligibly small. Let e_1 be the axial axis and e_2 be the circumferential axis in the wall. e_3 is radial axis normal to the cylindrical wall. The force exerted by the liquid on the wall

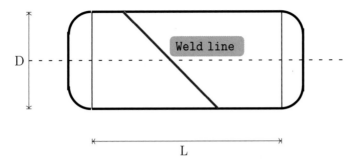

Fig. 3.3 Example 3.3.1

along its axial axis is $p(\pi R^2)$ and this force is balanced by the force resulted in the wall, which is $\sigma_{11}(2\pi R)$. Therefore the traction or stress σ_{11} is

$$p(\pi R^2) = \sigma_{11}(2\pi R)t$$

or

$$\sigma_{11} = \frac{pR}{2t}$$

Similarly for the stress perpendicular to the wall we have to balance the force exerted by the liquid on the wall and force resulted in the vessel's wall. That is

$$pRL = \sigma_{22}Lt$$

which produces

$$\sigma_{22} = \frac{pR}{t}$$

With no other stress inside the wall, the stress tensor can be written as

$$(\mathbf{S}) = \begin{pmatrix} pR/2t & 0 & 0 \\ 0 & pR/t & 0 \\ 0 & 0 & 0 \end{pmatrix}$$

In order to find the stress components in the weld beads, we need to do an orthogonal transformation of the stress tensor obtained previously. Basically, we rotate the current axis 45° clockwise to capture the weld line. The orthogonal or rotation tensor for this purpose is $\mathbf{Q}$,[3] which is

---

[3] Refer Appendix 1 for orthogonal transformation.

$$(\mathbf{Q}) = \begin{pmatrix} \cos(45) & \sin(45) & 0 \\ -\sin(45) & \cos(45) & 0 \\ 0 & 0 & 1 \end{pmatrix}$$

or

$$(\mathbf{Q}) = \begin{pmatrix} \sqrt{2}/2 & \sqrt{2}/2 & 0 \\ -\sqrt{2}/2 & \sqrt{2}/2 & 0 \\ 0 & 0 & 1 \end{pmatrix}$$

After the orthogonal transformation

$$\tilde{\mathbf{S}} = \mathbf{Q} \cdot \mathbf{S} \cdot \mathbf{Q}^T$$

we obtain

$$(\tilde{\mathbf{S}}) = \begin{pmatrix} 3pR/4t & pR/4t & 0 \\ pR/4t & 3pR/4t & 0 \\ 0 & 0 & 1 \end{pmatrix}$$

the same Cauchy stress tensor but written based on the rotated axis system. With the data provided the stress components in the weld line are

$$\tilde{\sigma}_{11} = 3pR/4t = 450 \text{ MPa}$$

$$\tilde{\sigma}_{22} = 3pR/4t = 450 \text{ MPa}$$

$$\tilde{\sigma}_{11} = pR/4t = 150 \text{ MPa}$$

Problem 3.3.1

Given is the Cauchy stress tensor $\mathbf{S}$ as follows:

$$(\mathbf{S}) = \begin{pmatrix} x_1 + x_2 & f(x_1, x_2) & 0 \\ f(x_1, x_2) & x_1 - 2x_2 & 0 \\ 0 & 0 & x_2 \end{pmatrix}$$

1. Determine the function $f(x_1, x_2)$ assuming that the body force is negligibly small and the origin is stress free.
2. Determine the stress vector at any point on the surface $x_1 = constant$ and $x_2 = constant$.

Problem 3.3.2

If $\mathbf{T}$ is any symmetrical tensor field, i.e. $\mathbf{T} = \mathbf{T}(x_i)$, $i = 1, 2, 3$, is then $\mathbf{S}$, with

$$\mathbf{S} = \nabla \times \nabla \times \mathbf{T}$$

statically allowable to represent the Cauchy stress tensor.

Problem 3.3.3

An observer $\mathcal{A}$ measures the Cauchy stress tensor at one particular point of a body and announce that his results are as expressed below

$$(\mathbf{S_A}) = \begin{pmatrix} 10 & 5 & 18 \\ 5 & 10 & 3 \\ 18 & 3 & 15 \end{pmatrix}$$

The observer $\mathcal{B}$ measures the Cauchy stress tensor at the same point using a different instrumentation and declares his result as

$$(\mathbf{S_B}) = \frac{1}{9} \begin{pmatrix} 226 & 35 & -104 \\ 35 & -26 & 41 \\ -104/ & 41 & 115 \end{pmatrix}$$

Do these observers measure the same state of stress at that point?

Problem 3.3.4

Determine the principal stresses and principal directions of the following Cauchy stress tensor,

$$(\mathbf{S}) = \begin{pmatrix} 5 & 10 & 8 \\ 10 & 7 & -6 \\ 8 & -6 & 12 \end{pmatrix}$$

Perform an orthogonal transformation of the above tensor using the following orthogonal matrix

$$(\mathbf{Q}) = \frac{1}{3} \begin{pmatrix} 1 & 2 & 2 \\ 2 & 1 & -2 \\ -2 & 2 & -1 \end{pmatrix}$$

and find the principal stresses and principal directions of this transformed tensor. Based on your answer what conclusion can you make?

Fig. 3.4 Initial
configuration

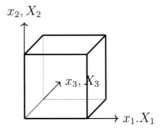

Problem 3.3.5

Show that the symmetrical Cauchy stress tensor **S** has a set of real eigenvalues and if
furthermore, they are distinct the associated principal directions form an orthonormal
basis system.

Problem 3.3.6

Figure 3.4 illustrates a rectangular bar under the axial loading and the components
of the resulting first Piola–Kirchhoff stress tensor are given below

$$
(\mathbf{T}) = \begin{pmatrix} \sigma & 0 & 0 \\ 0 & 0 & 0 \\ 0 & 0 & 0 \end{pmatrix}
$$

Calculate the components of the stress tensor after the body being stretched by $\alpha\%$
of its original length in axial direction x to the current configuration, as shown in
Fig. 3.5.

Fig. 3.5 After stretching

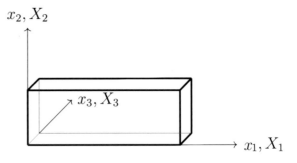

Fig. 3.6 First
Piola–Kirchhoff Stress
Tensor

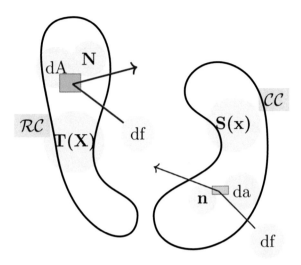

3.4 First and Second Piola–Kirchhoff Stress Tensor

In case of a body undergoes large deformation, such as the deformation of elastomers, rubber, or rubber-like materials or shape memory alloys, where the difference between undeformed and the deformed configuration is so significant, all infinite strain assumptions used previously are no longer valid.[4] Then there is a need to differentiate between the "true" stress and the "engineering" stress or in context of continuum mechanics, stress based on material Lagrangian coordinates and the stress based on spatial Eulerian coordinates. The latter has been discussed extensively in the previous section (3.1.3).

Figure 3.6 depicts the deformation process of a oriented surface area. The orientation of the surface area before deformation is represented by the normal vector **N** and the area dA. After deformation the orientation of this surface area changes to **n** with the magnitude of the infinitesimal area da. Since there is no change in loading the infinitesimal force vector **f** before and after the deformation is the same. In the current configuration **f** is given as

$$\mathbf{f} = \mathbf{S}^T \cdot \mathbf{n}\, da = \mathbf{S} \cdot \mathbf{n}\, da = \mathbf{S} \cdot d\mathbf{a}$$

while in the reference configuration the force vector takes the following form

$$\mathbf{f} = \mathbf{T}^T \cdot \mathbf{N}\, dA = \mathbf{T}^T \cdot d\mathbf{A}$$

---

[4] Large elastic deformation is also called geometrically nonlinear deformation while the plastic deformation is known as materially nonlinear deformation.

The tensor $\mathbf{T}$ is called first Piola–Kirchhoff stress tensor representing the state of stress in the body before deformation. Unlike the Cauchy stress tensor, first Piola–Kirchhoff is not symmetry. The relationship between these two stress tensors can be derived using the deformation gradient. From the previous chapter, we use initially this relation

$$d\mathbf{A} = J\mathbf{F}^T \cdot d\mathbf{a}$$

and obtain

$$\mathbf{S} \cdot d\mathbf{a} = \mathbf{T}^T \cdot d\mathbf{A}$$

Substitution of the above equation yields

$$\mathbf{S} \cdot d\mathbf{a} = \mathbf{T}^T \cdot d\mathbf{A} = \mathbf{T}^T \cdot J\mathbf{F}^T \cdot d\mathbf{a}$$

Thus, the relationship between these two stress tensors is

$$\mathbf{S} = J\mathbf{T}^T \cdot \mathbf{F}^T \tag{3.6}$$

Since $\mathbf{S}$ is symmetry, Eq. (3.6) can be rewritten as

$$\mathbf{S} = J\mathbf{F} \cdot \mathbf{T} \tag{3.7}$$

or the inversion of the equation (3.7) provides the expression of first Piola–Kirchhoff stress tensor in term of Cauchy stress tensor.

$$\mathbf{T} = (1/J)\mathbf{F}^{-1} \cdot \mathbf{S} \tag{3.8}$$

The nonsymmetry property of the first Piola–Kirchhoff stress tensor makes it unsuitable to be used in the construction of material equation. The modification has to be made in order to include state of stress in the reference configuration in the constitutive equation of material.

The idea behind this modification is to define another stress tensor to produce a force vector in the same manner as the first Piola–Kirchhoff tensor, i.e. via $\tilde{\mathbf{T}}^T \cdot d\mathbf{A}$ at the reference configuration. This tensor produces another force vector $\tilde{\mathbf{f}}$, which is fictive.

$$\tilde{\mathbf{T}}^T \cdot d\mathbf{A} = d\tilde{\mathbf{f}}$$

However, this fictive force vector can be related to the real force vector $\mathbf{f}$ by the deformation gradient through

$$\mathbf{f} = \mathbf{F} \cdot d\tilde{\mathbf{f}}$$

Substitution of previous equations into the above equation we have

$$\mathbf{f} = \mathbf{T}^T \cdot d\mathbf{A} = \mathbf{F} \cdot d\tilde{\mathbf{f}} = \mathbf{F} \cdot \tilde{\mathbf{T}}^T \cdot d\mathbf{A}$$

$$\mathbf{T}^T = \mathbf{F} \cdot \tilde{\mathbf{T}}^T$$

or we have

$$\tilde{\mathbf{T}}^T = \mathbf{F}^{-1} \cdot \mathbf{T}^T \tag{3.9}$$

We combine this equation with Eq. (3.6), which relates the Cauchy stress tensor with the first Piola–Kirchhoff tensor, and we get

$$\tilde{\mathbf{T}}^T = J\,\mathbf{F}^{-1} \cdot \mathbf{S} \cdot \mathbf{F}^{-T} \tag{3.10}$$

or otherwise,

$$\mathbf{S} = (1/J)\,\mathbf{F} \cdot \tilde{\mathbf{T}} \cdot \mathbf{F}^T \tag{3.11}$$

These three equations above relate the first and second Piola–Kirchhoff stress tensor and the second Piola–Kirchhoff tensor with the Cauchy stress tensor. One can clearly see that the second Piola–Kirchhoff stress tensor is symmetrical.

3.4.1 Examples and Problems

Example 3.4.1

Given are the deformation gradient $\mathbf{F}$ and the Cauchy stress tensor as follows:

$$(\mathbf{F}) = \begin{pmatrix} 0.94 & 0.31 & 0.23 \\ -0.210 & 1.12 & 0.11 \\ -0.32 & 0.11 & 1.12 \end{pmatrix}$$

$$(\mathbf{S}) = \begin{pmatrix} 5 & 10 & 8 \\ 10 & 7 & -6 \\ 8 & -6 & 12 \end{pmatrix}$$

Calculate the first Piola–Kirchhoff tensor and corresponding second Piola–Kirchhoff stress tensor.

Solution

The Jacobian J is calculated from the determinant of the deformation gradient

$$J = \det(\mathbf{F}) = \det \begin{pmatrix} 0.94 & 0.31 & 0.23 \\ -0.210 & 1.12 & 0.11 \\ -0.32 & 0.11 & 1.12 \end{pmatrix} = 1.30$$

The inverse of the deformation gradient **F** is readily calculated using MS Excel and obtained as

$$(\mathbf{F}^{-1}) = \begin{pmatrix} 0.96 & -0.25 & -0.17 \\ 0.15 & 0.86 & -0.12 \\ 0.26 & -0.16 & 0.86 \end{pmatrix}$$

Now as defined in Eq. (3.8) the first Piola–Kirchhoff tensor is

$$\mathbf{T} = (1/J)\mathbf{F}^{-1} \cdot \mathbf{S}$$

$$\mathbf{T} = (1/1.3) \begin{pmatrix} 0.96 & -0.25 & -0.17 \\ 0.15 & 0.86 & -0.12 \\ 0.26 & -0.16 & 0.86 \end{pmatrix} \cdot \begin{pmatrix} 5 & 10 & 8 \\ 10 & 7 & -6 \\ 8 & -6 & 12 \end{pmatrix}$$

$$= \begin{pmatrix} 0.29 & 2.85 & 2.26 \\ 2.72 & 2.66 & -1.72 \\ 2.12 & -1.17 & 4.27 \end{pmatrix}$$

Note that the first Piola–Kirchhoff is not symmetrical. The second Piola–Kirchhoff tensor can be obtained from Eq. (3.10).

$$\tilde{\mathbf{T}}^T = (1/J)\mathbf{F}^{-1} \cdot \mathbf{S} \cdot \mathbf{F}^{-T}$$

Basically

$$\tilde{\mathbf{T}}^T = \begin{pmatrix} 0.29 & 2.85 & 2.26 \\ 2.72 & 2.66 & -1.72 \\ 2.12 & -1.17 & 4.27 \end{pmatrix} \cdot \begin{pmatrix} 0.96 & 0.31 & 0.23 \\ -0.210 & 1.12 & 0.11 \\ -0.32 & 0.11 & 1.12 \end{pmatrix}^{-T}$$

$$\tilde{\mathbf{T}}^T = \begin{pmatrix} -0.82 & 2.23 & 1.58 \\ 2.23 & 2.91 & -1.18 \\ 1.58 & -1.18 & 4.380 \end{pmatrix}$$

Unlike the first Piola–Kirchhoff tensor, the second Piola–Kirchhoff tensor is symmetrical.

Problem 3.4.1

Based on the Cauchy stress tensor and the deformation gradient given in the **Example 1** above calculate the principal stresses and principal directions of the second Piola–Kirchhoff tensor and compare the results with the principal stresses and principal directions of the Cauchy stress tensor.

Problem 3.4.2

The relation between the Cauchy and the second Piola–Kirchhoff stress tensor is given in Eq. (3.10) or (3.11).

$$\mathbf{S} = (1/J)\,\mathbf{F} \cdot \tilde{\mathbf{T}} \cdot \mathbf{F}^T$$

Construct a tensor of fourth-order $\underline{\mathbf{F}}$ so that the above relation can be written as

$$\mathbf{S} = \underline{\mathbf{F}} \cdot\cdot\, \tilde{\mathbf{T}}$$

with "$\cdot\cdot$" is a double scalar product between tensors of higher order. Find also its inverse $\underline{\mathbf{F}}^{-1}$.

Problem 3.4.3

Given is a body with the motion

$$
\begin{aligned}
x_1 &= X_1 + (1/4)X_2^2 \\
x_2 &= (X_1 + 1)X_2^2 \\
x_3 &= X_3
\end{aligned}
\tag{3.12}
$$

The resulting Cauchy stress tensor $\mathbf{S}(X_i)$ as a function of material points $X_i, i = 1, 2, 3$ is determined as follows:

$$
(\mathbf{S}) = \begin{pmatrix}
X_1^2 & X_1 X_2 & 0 \\
X_1 X_2 & X_2^2 & X_2 X_3^2 \\
0 & X_2 X_3^2 & X_2^2
\end{pmatrix}
$$

Show that equations (3.12) represent a motion and calculate the first and second Piola–Kirchhoff tensor at the material point P $(1, 0.5, 1.5)$.

3.5   State of Stress in a Damaged Continuum

Remark: *The discussion in this section is made based on the classical works of Rabotnov, Murakami & Ohno and J. Betten. Unlike derivations in their papers, most of derivations made here has been modified from original indicial format to symbolic notation for simplification purposes. The indicial notations are only used in description of symmetry properties of the higher order tensors such as the continuity and damage tensors. Refer to Appendix I for exterior algebra and Levi-Civita symbols for further information*

In classical framework of continuum mechanics, material is postulated to be homogenous and uniform. In addition, material points are continuously and collectively attached to each other without any separation or boundary. Commonly, nonideal materials however exhibit some degree of defects and impurities. These defects and impurities appear in form of microcracks, for example, isotropic materials to become anisotropic one, which is completely different material. The evolution of damage in a material, for example, is due to the cohesion of microcracks and built-up of pores at the grain boundaries during the tertiary creep stage. This damage property is orientation dependent and directional. There is a need to modify our current understanding of the state of stress and the deformation of a body. Amongst the earliest researchers, Murakami & Ohno and Rabotnov introduced a tensor of second-order $\hat{\mathbf{S}}$ to describe the state of stress of a damaged continuum. This stress tensor is obtained from a linear transformation with a fourth-order tensor $\underline{\boldsymbol{\Theta}}$.

$$\hat{\mathbf{S}} = \underline{\boldsymbol{\Theta}} \cdot\cdot \, \mathbf{S} \longleftrightarrow \hat{\sigma}_{ij} = \Theta_{ijkl}\sigma_{kl}$$

The tensor $\hat{\mathbf{S}}$ is known as "net-tensor" of a damaged continuum. The aim of this section is to construct a damage tensor which is contained in the linear transformation $\underline{\boldsymbol{\Omega}}$ above.

Figures 3.5 and 3.6 below illustrate two tetrahedrons, both represent virgin or undamaged material and the damaged or defected one.

Depicted from Fig. 3.5, the normal vectors to all sides of tetrahedron are given as follows:

$$\mathbf{n}_1 = \mathbf{OC} \times \mathbf{OB} = |OC|\mathbf{e}_3 \times |OB|\mathbf{e}_2 = |OC||OB|\underline{\mathbf{E}} \cdot\cdot \, \mathbf{e}_2 \otimes \mathbf{e}_3$$

$$\mathbf{n}_2 = \mathbf{OA} \times \mathbf{OC} = |OA|\mathbf{e}_1 \times |OC|\mathbf{e}_3 = |OA||OC|\underline{\mathbf{E}} \cdot\cdot \, \mathbf{e}_3 \otimes \mathbf{e}_1$$

$$\mathbf{n}_3 = \mathbf{OB} \times \mathbf{OA} = |OB|\mathbf{e}_2 \times |OA|\mathbf{e}_1 = |OB||OA|\underline{\mathbf{E}} \cdot\cdot \, \mathbf{e}_1 \otimes \mathbf{e}_2$$

$$\mathbf{n} = \mathbf{BC} \times \mathbf{CA}$$

The sum of these normal vectors is equal to a zero vector

$$\mathbf{n}_1 + \mathbf{n}_2 + \mathbf{n}_3 + \mathbf{n} = 0$$

The same is the case for tetrahedron representing the damaged continuum. The normal vectors maintain the same direction but due to the defect have different length, i.e.

$$\mathbf{N}_1 = \alpha\mathbf{n}_1, \mathbf{N}_2 = \beta\mathbf{n}_2, \mathbf{N}_3 = \gamma\mathbf{n}_3, \mathbf{N} = \kappa\mathbf{n}$$

The parameter $\alpha < 1, \beta < 1, \gamma < 1$ and $\kappa < 1$ are numbers that describe the severity of the damaged surface area represented by these normal vectors. They are the ratio between net clean surface area over the original virgin area. Therefore the sum of the normal vectors of damaged continuum cannot equal to a zero vector.

$$\mathbf{N_1} + \mathbf{N_2} + \mathbf{N_3} + \mathbf{N} \neq \mathbf{0}$$

In case of isotropic damage, all these parameters are equal, i.e. $\alpha = \beta = \gamma = \kappa$ and it is clear that for the undamaged state $\alpha = \beta = \gamma = \kappa = 1$.

Furthermore, we consider Fig. 3.7 which shows the fictitious damaged configuration whereby all the normal vectors are embedded with damage parameter as the normal vectors of the damaged configuration, Fig. 3.6

$$\mathbf{m_1} = \mathbf{N_1}, \mathbf{m_2} = \mathbf{N_2}, \mathbf{m_3} = \mathbf{N_3}$$

except the fourth normal vector $\mathbf{m}$. The fictitious normal vector $\mathbf{m}$ has the same magnitude as the vector $\mathbf{N}$ of the damaged continuum but completely different from the normal vector of undamaged continuum $\mathbf{n}$. This vector is so adjusted so that the following condition valids.

$$\mathbf{m_1} + \mathbf{m_2} + \mathbf{m_3} + \mathbf{m} = \mathbf{0}$$

These two normal vectors are related by the following linear transformation

$$\mathbf{m} = \mathbf{\Psi} \cdot \mathbf{n}$$

Fig. 3.7 Virgin continuum

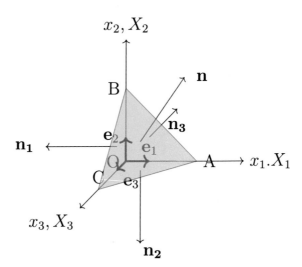

This linear transformation $\boldsymbol{\Psi}$ is a tensor second order containing all these damage parameters $\alpha < 1$, $\beta < 1$, and $\gamma < 1$. In terms of all normal vectors on the virgin configuration and on the fictitious damaged configuration the above equation can be written as

$$(\mathbf{m_1} + \mathbf{m_2} + \mathbf{m_3}) = \boldsymbol{\Psi} \cdot (\mathbf{n_1} + \mathbf{n_2} + \mathbf{n_3})$$

or

$$\boldsymbol{\Psi} \cdot (\mathbf{n_1} + \mathbf{n_2} + \mathbf{n_3}) = \alpha \mathbf{n_1} + \beta \mathbf{n_2} + \gamma \mathbf{n_3}$$

$$\boldsymbol{\Psi} \cdot (|OC||OB|\mathbf{e_3} \times \mathbf{e_2} + |OA||OC|\mathbf{e_1} \times \mathbf{e_3} + |OB||OA|\mathbf{e_2} \times \mathbf{e_1})$$

$$= \alpha|OC||OB|\mathbf{e_3} \times \mathbf{e_2} + \beta|OA||OC|\mathbf{e_1} \times \mathbf{e_3} + \gamma|OB||OA|\mathbf{e_2} \times \mathbf{e_1})$$

With the epsilon (or Levi- Civita) tensor of third-order $\underline{\mathbf{E}}$, thereby $\underline{\mathbf{E}} = \epsilon_{ijk}\mathbf{e_i} \otimes \mathbf{e_j} \otimes \mathbf{e_k}$. The symbol $\otimes$ represents the dyadic product of two vectors or tensors. Using this Levi-Civita tensor we introduce a concept of dual form of any vector or tensor.

Lemma 1 *Vector product of any two vectors can be represented by a double scalar product between a third-order epsilon or Levi-Civita tensor with a dyadic product constructed from the initial two vectors.*

$$\mathbf{c} = \mathbf{a} \times \mathbf{b} = \underline{\mathbf{E}} \cdot \cdot (\mathbf{b} \otimes \mathbf{a})$$

Lemma 2 *Any vector can be written as a double scalar product between Levi-Civita tensor and asymmetric tensor of second-order* $\mathbf{W}$*, i.e.*

$$2\mathbf{a} = \underline{\mathbf{E}} \cdot \cdot \mathbf{W}, \mathbf{W} = -\mathbf{W}^T$$

Lemma 3 *Any tensor of second order can be written as a double scalar product between Levi-Civita tensor and asymmetric tensor of third-order* $\underline{\mathbf{W}}$*, Asymmetry with respect to the last two indices.*

$$2\mathbf{A} = \underline{\mathbf{E}} \cdot \cdot \underline{\mathbf{W}},$$

With these lemmas, we are ready to continue further with the derivation of damage tensor. We proceed with the modification, that mentioned earlier in the Remark,

$$\boldsymbol{\Psi} \cdot (|OC||OB|\underline{\mathbf{E}} \cdot \cdot \mathbf{e_2} \otimes \mathbf{e_3} + |OA||OC|\underline{\mathbf{E}} \cdot \cdot \mathbf{e_3} \otimes \mathbf{e_1} + |OB||OA|\underline{\mathbf{E}} \cdot \cdot \mathbf{e_1} \otimes \mathbf{e_2})$$

$$= \alpha|OC||OB|\underline{\mathbf{E}} \cdot \cdot \mathbf{e_2} \otimes \mathbf{e_3} + \beta|OA||OC|\underline{\mathbf{E}} \cdot \cdot \mathbf{e_3} \otimes \mathbf{e_1} + \gamma|OB||OA|\underline{\mathbf{E}} \cdot \cdot \mathbf{e_1} \otimes \mathbf{e_2})$$

From the Fig. 3.7 all the side borders of the tetrahedron have the same length, therefore

Fig. 3.8 Damaged
continuum

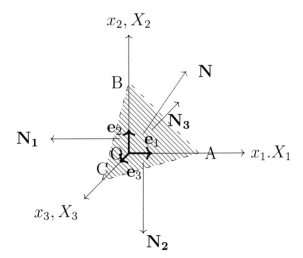

Fig. 3.9 Fictitious virgin
configuration

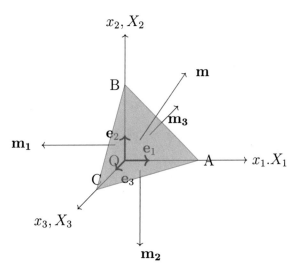

$$|OC||OB| = |OA||OB| = |OC||OA|$$

so that we can write

$$\boldsymbol{\Psi} \cdot (\underline{\mathbf{E}} \cdot \cdot \mathbf{e}_2 \otimes \mathbf{e}_3 + \underline{\mathbf{E}} \cdot \cdot \mathbf{e}_3 \otimes \mathbf{e}_1 + \underline{\mathbf{E}} \cdot \cdot \mathbf{e}_1 \otimes \mathbf{e}_2)$$

$$= \alpha \underline{\mathbf{E}} \cdot \cdot \mathbf{e}_2 \otimes \mathbf{e}_3 + \beta \underline{\mathbf{E}} \cdot \cdot \mathbf{e}_3 \otimes \mathbf{e}_1 + \gamma \underline{\mathbf{E}} \cdot \cdot \mathbf{e}_1 \otimes \mathbf{e}_2)$$

After factorization

$$\boldsymbol{\Psi} \cdot \underline{\mathbf{E}} \cdot\cdot (\mathbf{e}_2 \otimes \mathbf{e}_3 + \mathbf{e}_3 \otimes \mathbf{e}_1 + \mathbf{e}_1 \otimes \mathbf{e}_2)$$

$$= \alpha\underline{\mathbf{E}} \cdot\cdot \mathbf{e}_2 \otimes \mathbf{e}_3 + \beta\underline{\mathbf{E}} \cdot\cdot \mathbf{e}_3 \otimes \mathbf{e}_1 + \gamma\underline{\mathbf{E}} \cdot\cdot \mathbf{e}_1 \otimes \mathbf{e}_2)$$

The scalar product between the epsilon tensor $\underline{\mathbf{E}}$ and the linear transformation Ψ is a third-order tensor $\underline{\boldsymbol{\Psi}}$, so the above equation can be simplified as

$$\underline{\boldsymbol{\Psi}} \cdot\cdot (\mathbf{e}_2 \otimes \mathbf{e}_3 + \mathbf{e}_3 \otimes \mathbf{e}_1 + \mathbf{e}_1 \otimes \mathbf{e}_2)$$

$$= \alpha\underline{\mathbf{E}} \cdot\cdot \mathbf{e}_2 \otimes \mathbf{e}_3 + \beta\underline{\mathbf{E}} \cdot\cdot \mathbf{e}_3 \otimes \mathbf{e}_1 + \gamma\underline{\mathbf{E}} \cdot\cdot \mathbf{e}_1 \otimes \mathbf{e}_2) \tag{3.13}$$

By breaking up the dyadic products among the basis vector $\mathbf{e}_i, i = 1, 2, 3$ via the double scalar product we reduce the above equation to a simple vector equation, in which the component of third-order tensor $\underline{\boldsymbol{\Psi}}$ can be determined. Expanding these terms one by one, thereby we apply Einstein's convention, we have firstly,

$$\underline{\boldsymbol{\Psi}} \cdot\cdot \mathbf{e}_2 \otimes \mathbf{e}_3 = \Psi_{ijk}\mathbf{e}_i \otimes \mathbf{e}_j \otimes \mathbf{e}_k \cdot\cdot \mathbf{e}_2 \otimes \mathbf{e}_3$$

$$= \Psi_{ijk}\mathbf{e}_i\delta_{2k}\delta_{j3} = \Psi_{i32}\mathbf{e}_i$$

$$= \Psi_{132}\mathbf{e}_1 + \Psi_{232}\mathbf{e}_2 + \Psi_{332}\mathbf{e}_3$$

Putting all the terms together we have on the LHS of Eq. (3.11).

$$\Psi_{132}\mathbf{e}_1 + \Psi_{232}\mathbf{e}_2 + \Psi_{332}\mathbf{e}_3 + \Psi_{113}\mathbf{e}_1 + \Psi_{213}\mathbf{e}_2 + \Psi_{313}\mathbf{e}_3 + \Psi_{121}\mathbf{e}_1 + \Psi_{221}\mathbf{e}_2 + \Psi_{321}\mathbf{e}_3$$

With $\underline{\mathbf{E}} = \epsilon_{ijk}\mathbf{e}_i \otimes \mathbf{e}_j \otimes \mathbf{e}_k$ we have on the RHS of the equation (3.11) the following expression

$$= \alpha\epsilon_{132}\mathbf{e}_1 + \alpha\epsilon_{232}\mathbf{e}_2 + \alpha\epsilon_{332}\mathbf{e}_3 + \beta\epsilon_{132}\mathbf{e}_1 + \beta\epsilon_{232}\mathbf{e}_2 + \beta\epsilon_{332}\mathbf{e}_3$$

$$+ \gamma\epsilon_{132}\mathbf{e}_1 + \gamma\epsilon_{232}\mathbf{e}_2 + \gamma\epsilon_{332}\mathbf{e}_3$$

By definition of epsilon or Levi- Civita tensor, it takes value of +1 for even permutation of indices and -1 for odd permutation, otherwise it takes the value of zero. After componentwise comparison, we conclude

$$\Psi_{123} = -\Psi_{132} = \alpha$$

$$\Psi_{231} = -\Psi_{213} = \beta$$

$$\Psi_{312} = -\Psi_{321} = \gamma$$

The other components are zeros. The third-order tensor $\Psi_{i[jk]}$ is asymmetric with respect to the last two indices as indicated by square bracket. The equivalent second-order tensor to the tensor Ψ_{ijk} can be obtained from the dual form discussed earlier. In this case, we have

$$2\Psi = \underline{\mathbf{E}} \cdot \cdot \underline{\Psi} \longleftrightarrow 2\Psi_{ij} = \epsilon_{ilk} \Psi_{jkl}$$

and after manipulating the above dual form we have the continuity tensor of second-order Ψ

$$(\Psi) = \begin{pmatrix} \alpha & 0 & 0 \\ 0 & \beta & 0 \\ 0 & 0 & \gamma \end{pmatrix}$$

The damage tensor Ω is then defined as

$$\Omega = \mathbf{I} - \Psi \tag{3.14}$$

The tensor $\mathbf{I}$ is the identity tensor of second order. Based on the same traction vector obtained from the Cauchy stress tensor of undamaged continuum, $\mathbf{S}$, and the net-stress tensor, $\hat{\mathbf{S}}$, of a damaged continuum the relationship between these two continua can be derived. In short, they are related by the following transformation.

$$\hat{\mathbf{S}} = \Psi^{-1} \cdot \mathbf{S} \tag{3.15}$$

3.5.1 Examples and Problems

Example 3.5.1

Show that the net-stress tensor

$$\hat{\mathbf{S}} = \Psi^{-1} \cdot \mathbf{S}$$

is not symmetric and derive the symmetrical and asymmetrical parts of the net-stress tensor $\hat{\mathbf{S}}$.

Solution

The Cauchy stress tensor $\mathbf{S}$ and the continuity tensor Ψ are symmetrical. From the equation

$$\mathbf{S} = \Psi \cdot \hat{\mathbf{S}}$$

we get

$$\mathbf{S} = \Psi \cdot \hat{\mathbf{S}} = \mathbf{S}^T = (\Psi \cdot \hat{\mathbf{S}})^T = \hat{\mathbf{S}}^T \cdot \Psi^T$$

or

$$\boldsymbol{\Psi} \cdot \hat{\mathbf{S}} = \hat{\mathbf{S}}^T \cdot \boldsymbol{\Psi}$$

$$\hat{\mathbf{S}} = \boldsymbol{\Psi}^{-1} \cdot \hat{\mathbf{S}}^T \cdot \boldsymbol{\Psi}$$

which shows that the net-stress tensor $\hat{\mathbf{S}}$ is not symmetrical. The transposition of the net-stress tensor $\hat{\mathbf{S}}$ is

$$\hat{\mathbf{S}}^T = (\boldsymbol{\Psi}^{-1} \cdot \hat{\mathbf{S}}^T \cdot \boldsymbol{\Psi})^T = \boldsymbol{\Psi}^T \cdot \hat{\mathbf{S}} \cdot \boldsymbol{\Psi}^{-T} = \boldsymbol{\Psi} \cdot \hat{\mathbf{S}} \cdot \boldsymbol{\Psi}^{-1}$$

The symmetrical part of the net-stress tensor is

$$Sym\,\hat{\mathbf{S}} = \frac{1}{2}(\boldsymbol{\Psi}^{-1} \cdot \hat{\mathbf{S}}^T \cdot \boldsymbol{\Psi} + \boldsymbol{\Psi} \cdot \hat{\mathbf{S}} \cdot \boldsymbol{\Psi}^{-1})$$

and its asymmetrical counterpart is

$$Assym\,\hat{\mathbf{S}} = \frac{1}{2}(\boldsymbol{\Psi}^{-1} \cdot \hat{\mathbf{S}}^T \cdot \boldsymbol{\Psi} - \boldsymbol{\Psi} \cdot \hat{\mathbf{S}} \cdot \boldsymbol{\Psi}^{-1})$$

Problem 3.5.1

Derive the equation of motion of a body with damaged continuum, assuming the body force is negligibly small. The damage tensor is given as

$$\boldsymbol{\Omega} = \begin{pmatrix} 1 - \alpha & 0 & 0 \\ 0 & 1 - \beta & 0 \\ 0 & 0 & 1 - \gamma \end{pmatrix}$$

Whereby the damage parameters $1 - \alpha$, $1 - \beta$ and $1 - \gamma$ take the values between 0 and 1.

Chapter 4
Constitutive Relations

The aim of this chapter is to provide a brief drive-through a vast area of mathe-matical modelling of material properties in continuum mechanics. The readers are encouraged to seek further knowledge from other sources listed in the reference.

Experience taught us that a body responses differently to the load applied to it. The way of body rotates, translates or deforms depends solely on the type of material the body constitutes. If we have the equation of motion, a topic which we will discuss thoroughly next chapter, we are still not able to determine the motion. The missing link is here the information about the material properties, or precisely the equation that relates the state of stress in a body and the associated deformation field. These equations are known as constitutive relations or material equations. Material equations describe the rheological behaviour of materials in connexion between stresses and strain or strain rate in a body under consideration.

In formulating the material equation for the material property, one has to make sure that this formulation fulfils the requirement of the principle of material frame-indifferent or also known as the principle of material objectivity. The relationship between state of stress and deformation or deformation rate as formulated by the material equation must be able to describe the same state of stress in the body independently whether the observer is moving or not. In other words, the derived material equation must be objective or form-invariant against rigid body rotation or translation of the observer. Another aspect of material property that is to be taken into account in formulating material equation is the principle of material symmetry or simply material symmetry. The term "material symmetry" is referred to the property of material that is dependent on the direction of observation. The easiest example of this direction dependent property is a piece wood. For the same loading, the responses of a piece of wood are different in radial, axial and circumferential direction. If this piece of wood happened to give the same response in radial direction then it is called radial isotropy otherwise it is anisotropic material. In contrary if the response to the external loading is the same in all directions then the material is said to be isotropic. Usually, the property of material symmetry is manifested in the number of material parameters required to cater those behaviours. For an isotropic linear-elastic material, there are only two material constants required to describe completely material response but 9 material constants are needed to describe the response of an orthotropic linear-elastic material completely.

It is also another challenging task to formulate the material equation if the material properties evolve during the same loading process. The formation of damage as discussed in the previous chapter or the nonlinearities of material properties, such as change of material property from elastic to plastic material during the torsion loading are few examples to mention. There are materials whose properties depend on the type of loading or rate of loading. Why, for example, toothpaste behaves like solid but is completely fluid under loading? Why some plasticine clays bounce back like a rubber ball when thrown hardly to the floor but deform plastically if pressed slowly? Another aspect to be considered in formulation of material equation is the ageing property of the materials. Biological tissues are example of material with ageing property. These are questions to be answered by the future researchers in this area. In the following section we discuss.

4.1 Elastic Materials

Phenomenologically an elastic materialis understood as deformed material that returns immediately and completely to its original configuration—shape and volume—upon releasing the load. In the rational mechanics, which popularized by Clifford Truesdell, elastic material is classified as spontaneous material, i.e. its current state of stress depends solely upon the actual state of strain experienced. In general term, elastic material can be represented by the following equation

$$\mathbf{S} = \mathcal{F}(\mathbf{F}) \longleftrightarrow \sigma_{ij} = \mathcal{F}_{ij}(F_{pq}) = \mathcal{F}_{ij}\left(\frac{\partial x_p}{\partial X_q}\right) \tag{4.1}$$

The Cauchy stress in material is expressed as a symmetrical tensor-valued function of the deformation gradient $\mathbf{F}$.

Now we examine if the above equation is objective or form-invariant. Any tensor-valued function $\mathbf{B}$ of tensor argument $\mathbf{A}$

$$\mathbf{B} = \mathcal{F}(\mathbf{A})$$

is said to be form-invariant or objective against rigid body rotation $\mathbf{Q}$ if

$$\mathbf{Q} \cdot \mathbf{B} \cdot \mathbf{Q}^T = \mathcal{F}(\mathbf{Q} \cdot \mathbf{A} \cdot \mathbf{Q}^T) \tag{4.2}$$

Assuming we have two observers formulating the material equation for elastic material as shown by Eq. (4.1). The difference between these observers is that one is moving while the other one is stationary describing the state of stress in a body. As

for rigid body rotation and translation, the both observers are related by the following
equation of rigid body transformation,

$$\hat{\mathbf{x}} = \mathbf{Q} \cdot \mathbf{x} + \mathbf{c(t)} \tag{4.3}$$

$$\hat{\mathbf{x}} = \mathbf{Q} \cdot \mathbf{x} + \mathbf{c(t)}$$

with $\hat{\mathbf{x}}$ the position vector of moving observer. From the moving observer, the material
equation takes the following form,

$$\hat{\mathbf{S}} = \mathcal{F}(\hat{\mathbf{F}}) \tag{4.4}$$

Furthermore because of Eq. (4.2), the deformation gradient $\hat{\mathbf{F}}$ for a moving
observer has the form

$$\hat{\mathbf{F}} = \mathbf{Q} \cdot \mathbf{F}$$

Referring to the objectivity condition stated in Eq. (4.2), the material equation formulated by the moving observer is

$$\hat{\mathbf{S}} = \mathbf{Q} \cdot \mathbf{S} \cdot \mathbf{Q}^T = \mathcal{F}(\mathbf{Q} \cdot \mathbf{F}) \tag{4.5}$$

So the material equation (4.5) with the deformation gradient $\mathbf{F}$ as argument does
not fulfil the objectivity condition and therefore is not objective or form-invariant.
However instead of using the deformation gradient itself but the left stretch tensor $\mathbf{V}$
from the polar decomposition of the deformation gradient $\mathbf{F} = \mathbf{V} \cdot \mathbf{R}$. By inserting
$\mathbf{F} = \mathbf{V} \cdot \mathbf{R}$ and $\mathbf{R} = \mathbf{Q}^T$ we have

$$\hat{\mathbf{S}} = \mathbf{Q} \cdot \mathbf{S} \cdot \mathbf{Q}^T = \mathcal{F}(\mathbf{Q} \cdot \mathbf{V} \cdot \mathbf{Q}^T) \tag{4.6}$$

The objectivity of the formulated material equation is secured if we use the left
stretch tensor to represent the deformation of the body.

$$\mathbf{S} = \mathcal{F}(\mathbf{V})$$

Applying the Cayley–Hamilton theorem from the theory of isotropic functions, the
tensor-valued tensor function $\mathcal{F}$ can be expressed as

$$\mathbf{S} = \phi_0 \mathbf{I} + \phi_1 \mathbf{V} + \phi_2 \mathbf{V}^2 \tag{4.7}$$

Thereby ϕ_0, ϕ_1, ϕ_2 are scalar-valued functions of irreducible invariants of the left
stretch tensor $\mathbf{V}$, which can be expressed in terms of elastic constants. Eq. (4.7)
in general describes the behaviour of isotropic elastic material for finite or large
deformation. This nonlinearity is characterized by $\phi_2 \neq 0$. Rubber or rubber-like
elasticity can be modelled by this equation.

4.1.1   Elasticity Tensor

The constitutive equation for linear-elastic material is commonly given as

$$\mathbf{S} = \underline{\mathbf{E}} : \hat{\mathbf{E}} \longleftrightarrow \sigma_{ij} = E_{ijkl}\epsilon_{kl} \tag{4.8}$$

with $\hat{\mathbf{E}}$ is the classical strain tensor or precisely the linear Lagrangian strain tensor defined in Eq. (2.17).[1] The tensor $\underline{\mathbf{E}}$ is a fourth-order tensor containing elasticity constants. Basically, the above material equation is a generalization of a one-dimensional Hooke's law. Mathematically, Eq. (4.8) can be interpreted as a linear transformation between the Cauchy stress tensor and infinitesimal strain tensor. It is then a linear relation in context of material properties.[2]

The objectivity of the material equation (4.8) is secured by the transformation properties of the components of the elasticity tensors. Let $\tilde{E}_{ijkl}$ be the element of elasticity tensor registered by a moving observer, then the relationship between the same elasticity tensor observed by a stationary observer is given by

$$\tilde{E}_{ijkl} = Q_{ip}Q_{jq}Q_{kr}Q_{ls}E_{pqrs}$$

Thereby Q_{kl} is an orthogonal transformation tensor with $Q_{pq}Q_{rq} = \delta_{pr}$.

4.1.2   Elastic Potential

Consider the force versus extension behaviour of a simple linear spring subjected to uniaxial loading. The potential energy stored in the spring is given by

$$\psi = \frac{1}{2}k\epsilon^2 \tag{4.9}$$

We obtain the well-known Hooke's law by differentiating the elastic potential energy with respect to the extension ϵ. Further differentiation of the potential energy function delivers the material or spring constant itself.

$$\frac{d\psi}{d\epsilon} = k\epsilon = f$$

$$\frac{d^2\psi}{d\epsilon^2} = k$$

---

[1] For small infinitesimal deformation it is insignificant to differentiate between strain tensor $\hat{\mathbf{E}}$ or left stretch tensor $\mathbf{V}$.

[2] This linear relation is called the physical linearity.

The generalization of the above potential energy function for three-dimensional elastic continuum is as follows:

$$\psi = \frac{1}{2}\hat{\mathbf{E}} : \underline{\mathbf{E}} : \hat{\mathbf{E}} \longleftrightarrow \psi = \frac{1}{2}E_{ijkl}\epsilon_{ij}\epsilon_{kl} \tag{4.10}$$

In continuum mechanics, this potential ψ is called elastic potential or strain energy density function. Similarly twice differentiation of the strain energy function ψ with respect to strain tensor ϵ_{ij} and ϵ_{kl} yields

$$E_{ijkl} = \frac{\partial^2 \psi}{\partial \epsilon_{ij} \partial \epsilon_{kl}} = \frac{\partial^2 \psi}{\partial \epsilon_{kl} \partial \epsilon_{ij}} = E_{klij}$$

It is visible that the elasticity tensor is symmetrical with respect to the first and second index pairs. This symmetry is known as major symmetry. Another type of symmetry is a minor symmetry, i.e. symmetry with respect to first two indices and the last two indices,

$$E_{ijkl} = E_{jikl}, \; E_{ijkl} = E_{ijlk}$$

These two minor symmetries resulted from the constitutive equation (4.8) with the symmetry of the Cauchy stress tensor and Lagrangian strain tensor.

Due to these symmetries, minor and major the elasticity tensor has only 21 independent elements out of the initial 81 components.

4.1.3 Linear Isotropic Elastic Material

The isotropy requirement on material elastic tensor leads to the fulfilment of the objectivity condition automatically independent of the rotation of moving observer, i.e. $\tilde{E}_{ijkl} = E_{pqrs}$ always, regardless the direction of moving observer chooses. This requirement necessitates the elasticity tensor to take a form of a linear combination of dyadic products of unity tensors. Written in indicial notation there are only 3 permutations of a such independent linear combinations

$$\delta_{ij}\delta_{kl}, \; \delta_{il}\delta_{kj}, \; \delta_{ik}\delta_{jl}$$

Thus

$$E_{pqrs} = \lambda\delta_{pq}\delta_{rs} + \mu\delta_{pr}\delta_{qs} + \kappa\delta_{ps}\delta_{qr}$$

Furthermore due to symmetry, minor and major symmetry discussed previously, $\mu = \kappa$ so that the following equation

$$E_{pqrs} = \lambda\delta_{pq}\delta_{rs} + \mu(\delta_{pr}\delta_{qs} + \delta_{ps}\delta_{qr}) \tag{4.11}$$

represents the material equation for linear isotropic elastic material. The material constants λ, and μ are called Lamè constants and there are closely related to elastic (Young) modulus, E, bulk modulus G and the Poisson ratio ν. The solved problems at the end of this section will discuss this relationship.

With Eq. (4.11) we are able now to produce the relationship between Cauchy stress tensor and strain tensor for small deformation of a linear isotropic elastic material

$$\mathbf{S} = \underline{\mathbf{E}} : \hat{\mathbf{E}} \longleftrightarrow \sigma_{ij} = (\lambda \delta_{pq} \delta_{rs} + \mu(\delta_{pr}\delta_{qs} + \delta_{ps}\delta_{qr}))\epsilon_{rs}$$

$$\sigma_{ij} = \lambda \delta_{pq} \delta_{rs} \epsilon_{rs} + \mu(\delta_{pr}\delta_{qs}\epsilon_{rs} + \delta_{ps}\delta_{qr}\epsilon_{rs})$$

$$\sigma_{pq} = \lambda \delta_{pq} \epsilon_{ss} + \mu(\epsilon_{pq} + \epsilon_{qp})$$

$$\sigma_{pq} = \lambda \delta_{pq} \epsilon_{ss} + 2\mu(\epsilon_{pq} + \epsilon_{qp}) \longleftrightarrow \mathbf{S} = \lambda \, tr(\hat{\mathbf{E}})\mathbf{I} + 2\mu\hat{\mathbf{E}} \qquad (4.12)$$

The above equation is known as the Lame–Navier equation in elasticity. From two fundamental experiments, torsion and uniaxial tensile testing, the Lamè constants can be determined. They are

$$\lambda = \frac{\nu E}{(1 - \nu)(1 - 2\nu)} \qquad (4.13)$$

$$\mu = G = \frac{E}{2(1 + \nu)}$$

Further important relations are

$$E = \frac{\nu(3\lambda + 2\nu)}{(\lambda + \mu)}$$

$$\nu = \frac{\lambda}{2(\lambda + \mu)}$$

4.1.4 Linear Orthotropic Elastic Material

In general, orthotropy is a class of anisotropic material that possess three mutually orthogonal planes of reflection symmetry. The reflection planes are $(\mathbf{1}, \mathbf{2})$, $(\mathbf{1}, \mathbf{3})$ and $(\mathbf{3}, \mathbf{2})$ as shown in Fig. 6.6. Therefore the reflection axes are x_1, x_2 and x_3 with corresponding orthogonal tensors of reflection $\mathbf{Q}^3$

---

[3] The reflection tensor is an orthogonal tensor with determinant equals -1.

Fig. 4.1 Three orthogonal planes

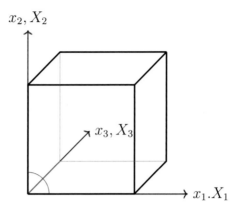

$$(\mathbf{Q}) = \begin{pmatrix} -1 & 0 & 0 \\ 0 & 1 & 0 \\ 0 & 0 & 1 \end{pmatrix}$$

$$(\mathbf{Q}) = \begin{pmatrix} 1 & 0 & 0 \\ 0 & -1 & 0 \\ 0 & 0 & 1 \end{pmatrix}$$

$$(\mathbf{Q}) = \begin{pmatrix} 1 & 0 & 0 \\ 0 & 1 & 0 \\ 0 & 0 & -1 \end{pmatrix}$$

Those are the reflection tensors about an orthotropic axis, **1**, **2** and **3**, respectively. So the two observers describing the material equation are distinguished from each other by this reflection tensor. For the first observer, the material equation is written as (Fig. 4.1)

$$\sigma_{ij} = E_{ijkl}\epsilon_{kl}$$

whilst for the second or moving observer the material equation for the same material takes the following form

$$\hat{\sigma}_{ij} = \hat{E}_{ijkl}\hat{\epsilon}_{kl}$$

furthermore the relationship between these quantities is listed as below, the stress and strain transformation

$$\hat{\sigma}_{ij} = Q_{ik}Q_{jl}\sigma_{kl}$$

$$\hat{\epsilon}_{ij} = Q_{ik}Q_{jl}\epsilon_{kl}$$

and the material tensor

$$\hat{E}_{ijkl} = Q_{ip} Q_{jq} Q_{kr} Q_{ls} E_{pqrs}$$

For the orthotropic conditions $\hat{E}_{ijkl} = E_{pqrs}$ to be fulfilled for all reflection tensors listed above, the material tensor must have the following structure with nine independent material constants.[4]

$$E_{ijkl} = \begin{pmatrix} E_{1111} & E_{1122} & E_{1133} & 0 & 0 & 0 \\ & E_{2222} & E_{2233} & 0 & 0 & 0 \\ & & E_{3333} & 0 & 0 & 0 \\ & & & E_{2323} & 0 & 0 \\ & symm & & & E_{1313} & 0 \\ & & & & & E_{1212} \end{pmatrix}$$

4.1.5   Examples and Problems

Example 4.1.1

Find the inversion of the material equation (4.12).

Solution

Tee double scalar product Eq. (4.12) delivers the trace of the linearized Lagrangian strain tensor $\hat{\mathbf{E}}$,

$$\mathbf{I} \cdot \cdot \mathbf{S} = tr(\mathbf{S}) = \mathbf{I} \cdot \cdot \lambda \, tr(\hat{\mathbf{E}})\mathbf{I} + 2\mu \mathbf{I} \cdot \cdot \hat{\mathbf{E}}$$

With $\mathbf{I} \cdot \cdot \mathbf{I} = 3$, above equation reduces to

$$\mathbf{I} \cdot \cdot \mathbf{S} = tr(\mathbf{S}) = (3\lambda + 2\mu) \, tr(\hat{\mathbf{E}})$$

From Eq. (4.12), we have then

$$\hat{\mathbf{E}} = \frac{1}{2\mu}(\mathbf{S} - \lambda \, tr(\hat{\mathbf{E}})\mathbf{I} = \frac{1}{2\mu}\left(\mathbf{S} - \frac{\lambda}{3\lambda + 2\mu} tr(\mathbf{S})\mathbf{I}\right)$$

Problem 4.1.2

Verify the relationship between the Lamè constants and the experimentally determined elastic constants, E, G and ν as described by Eq. (4.13).

---

[4] Refer Appendix for the Voigt Notation and complete derivation of orthotropic structure of material tensor.

4.2 Hyperelasticity

In context of continuum mechanics or rational theory of material equations elasticity is subset of hyperelasticity. Hyperelastic materials are materials whose stress-strain behaviour can be derived from stored potential energy, or more specific, the strain energy density function. Hyperelastic modelling of material property is a energy-based modelling rather than force-based modelling, as we see in generalization of classical Hooke's law for linear isotropic elastic material modelling.

Main reason why some rheologists and material researchers migrate from force-based formulation such as classical generalization of Hooke's law is its inability to capture the observed material behaviour undergoing large deformation whilst maintaining the elastic properties. Rubber or rubber-like elasticity are among common examples of such material behaviour. Soft tissues or biological tissues for instance show this characteristics. They can undergo large deformation yet able to return to initial configuration. The models are also efficient in numerical implementation and also accessible in major FE softwares.

The key characteristics in categorizing hyperelastic material is the use of the strain energy density function or the strain energy function ψ in establishing its constitutive equation. It is a scalar-valued tensor function of right Cauchy–Green deformation tensor $\mathbf{C}$ in material or Lagrangian representation or of left Cauchy–Green deformation tensor $\mathbf{B}$ in spatial or Eulerian representation. General constitutive equation for hyperelastic material in Lagrangian formulation with the second Piola–Kirchhoff stress tensor $\tilde{\mathbf{T}}$ is as follows:

$$\tilde{\mathbf{T}} = 2\frac{\partial \psi(\mathbf{C})}{\partial \mathbf{C}} \tag{4.14}$$

or in Eulerian formulation with Cauchy stress tensor $\mathbf{S}$

$$\mathbf{S} = 2J^{-1}\mathbf{B} \cdot \frac{\partial \psi(\mathbf{B})}{\partial \mathbf{B}} \tag{4.15}$$

In case of isotropic hyperelastic material, the strain energy function ψ can be expressed as a scalar function of principal invariants of the right Cauchy–Green deformation tensor or the left Cauchy–Green tensor. The invariants of both deformation tensors are the same. Therefore, we have

$$\psi(\mathbf{C}) = \psi(I_{\mathbf{C}}, II_{\mathbf{C}}, III_{\mathbf{C}}) \tag{4.16}$$

or

$$\psi(\mathbf{C}) = \psi(I_{\mathbf{B}}, II_{\mathbf{B}}, III_{\mathbf{B}})$$

In the following, we focus on the derivation of the constitutive equation of isotropic hyperelastic material in material or Lagrangian description. We establish the relationship between the second Piola–Kirchhoff $\tilde{\mathbf{T}}$ and the differential of the strain energy function with respect to the right Cauchy–Green deformation tensor $\mathbf{C}$. The derivation of material equation in spatial or Eulerian description can be done similarly. From Eq. (4.16), we obtain after the differentiation according to Eq. (4.14)

$$2\frac{\partial \psi(\mathbf{C})}{\partial \mathbf{C}} = 2\frac{\partial \psi(I_{\mathbf{C}}, II_{\mathbf{C}}, III_{\mathbf{C}})}{\partial \mathbf{C}}$$

Applying the chain rule, we have the following,[5]

$$\frac{\partial \psi}{\partial \mathbf{C}} = \frac{\partial \psi}{\partial I_{\mathbf{C}}} \cdot \frac{\partial I_{\mathbf{C}}}{\partial \mathbf{C}} + \frac{\partial \psi}{\partial II_{\mathbf{C}}} \cdot \frac{\partial II_{\mathbf{C}}}{\partial \mathbf{C}} + \frac{\partial \psi}{\partial III_{\mathbf{C}}} \cdot \frac{\partial III_{\mathbf{C}}}{\partial \mathbf{C}}$$

$$\frac{\partial \psi}{\partial \mathbf{C}} = \left(\frac{\partial \psi}{\partial I_{\mathbf{C}}} + I_{\mathbf{C}}\frac{\partial \psi}{\partial II_{\mathbf{C}}}\right)\mathbf{I} - \frac{\partial \psi}{\partial II_{\mathbf{C}}}\mathbf{C} + \frac{\partial \psi}{\partial III_{\mathbf{C}}}\mathbf{C}^{-1} \qquad (4.17)$$

or in short form

$$\frac{\partial \psi}{\partial \mathbf{C}} = \beta_0\mathbf{I} + \beta_1\mathbf{C} + \beta_2\mathbf{C}^{-1}$$

With the coefficients

$$\beta_0 = \frac{\partial \psi}{\partial I_{\mathbf{C}}} + I_{\mathbf{C}}\frac{\partial \psi}{\partial II_{\mathbf{C}}}$$

$$\beta_1 = -\frac{\partial \psi}{\partial II_{\mathbf{C}}}$$

and

$$\beta_2 = III_{\mathbf{C}}\frac{\partial \psi}{\partial III_{\mathbf{C}}}$$

According to the Cayley–Hamilton theorem, the tensor $\mathbf{C}^{-1}$ can be expressed as a linear combination of $\mathbf{I}$, $\mathbf{C}$ and $\mathbf{C}^2$, so that we can rewrite as

$$\frac{\partial \psi}{\partial \mathbf{C}} = \alpha_0\mathbf{I} + \alpha_1\mathbf{C} + \alpha_2\mathbf{C}^2$$

with new coefficients

$$\alpha_0 = \frac{\partial \psi}{\partial I_{\mathbf{C}}} + I_{\mathbf{C}}\frac{\partial \psi}{\partial II_{\mathbf{C}}} + II_{\mathbf{C}}\frac{\partial \psi}{\partial III_{\mathbf{C}}}$$

$$\alpha_1 = -\frac{\partial \psi}{\partial II_{\mathbf{C}}} - I_{\mathbf{C}}\frac{\partial \psi}{\partial III_{\mathbf{C}}}$$

and

---

[5] Refer to the Gâteaux variation in the appendix for the derivative of the invariant of a tensor with respect to a tensor.

$$\alpha_2 = \frac{\partial \psi}{\partial III_C}$$

Now the material description of the constitutive equation for isotropic hyperelastic material takes the following form

$$\tilde{\mathbf{T}} = 2\frac{\partial \psi}{\partial \mathbf{C}} = 2\alpha_0 \mathbf{I} + 2\alpha_1 \mathbf{C} + 2\alpha_2 \mathbf{C}^2 \tag{4.18}$$

or

$$\tilde{\mathbf{T}} = 2\frac{\partial \psi}{\partial \mathbf{C}} = \beta_0 \mathbf{I} + \beta_1 \mathbf{C} + \beta_2 \mathbf{C}^{-1} \tag{4.19}$$

In case of the strain energy function is expressed as a scalar function of the invariants of the left Cauchy–Green deformation tensor $\mathbf{B}$, then the constitutive equation obtained gives the relationship between Cauchy stress tensor $\mathbf{S}$ and the linear combination of left Cauchy–Green deformation tensor

$$\mathbf{S} = 2J^{-1}\mathbf{B}\frac{\partial \psi}{\partial \mathbf{B}} = \alpha_0 \mathbf{I} + \alpha_1 \mathbf{B} + \alpha_2 \mathbf{B}^2 \tag{4.20}$$

$$\mathbf{S} = 2J^{-1}\mathbf{B}\frac{\partial \psi}{\partial \mathbf{B}} = \beta_0 \mathbf{I} + \beta_1 \mathbf{B} + \beta_2 \mathbf{B}^{-1} \tag{4.21}$$

with new coefficients

$$\alpha_0 = 2J^{-1} III_{\mathbf{B}}\frac{\partial \psi}{\partial III_{\mathbf{B}}}$$

$$\alpha_1 = 2J^{-1}\left(\frac{\partial \psi}{\partial I_{\mathbf{B}}} + I_{\mathbf{B}}\frac{\partial \psi}{\partial II_{\mathbf{B}}}\right)$$

$$\alpha_2 = -2J^{-1}\left(\frac{\partial \psi}{\partial II_{\mathbf{B}}}\right)$$

With the coefficients

$$\beta_0 = 2J^{-1}\left(II_{\mathbf{B}}\frac{\partial \psi}{\partial II_{\mathbf{B}}} + III_{\mathbf{B}}\frac{\partial \psi}{\partial III_{\mathbf{B}}}\right)$$

$$\beta_1 = 2J^{-1}\frac{\partial \psi}{\partial I_{\mathbf{B}}}$$

and

$$\beta_2 = -2J^{-1}III_{\mathbf{B}}\frac{\partial \psi}{\partial II_{\mathbf{B}}}$$

There are several hyperelastic models are available and commonly used. They are different from each other only by different definitions of strain energy density function ψ. Amongst others are Mooney–Rivlin material, St. Venant–Kirchhoff and Ogden Materials. In this manuscript, we constrain the scope to Mooney–Rivlin and Ogden materials.

4.2.1 Mooney–Rivlin Materials

The strain density function ψ for Mooney–Rivlin material is defined as a linear combination of the first and second invariants of the left Cauchy–Green deformation tensor $\mathbf{B}$ and $\mathbf{B} = \mathbf{F} \cdot \mathbf{F}^T$.

$$\psi = c_1 J^{-2/3}(I_1 - 3) + c_2 J^{-4/3}(I_2 - 3) + (1/D)(J - 1)^2 \qquad (4.22)$$

With J is the Jacobian, $J = \det \mathbf{F}$ and I_1, I_2 are the first and second invariant of the tensor $\mathbf{B}$. If the eigenvalues of the deformation gradient $\mathbf{F}$ are λ_1, λ_2 and λ_3 then Eq. (4.22) can be rewritten as

$$\psi = c_1 (\lambda_1 \lambda_2 \lambda_3)^{-2/3}(I_1 - 3) + c_2 (\lambda_1 \lambda_2 \lambda_3)^{-4/3}(I_2 - 3) \qquad (4.23)$$

with

$$J = \lambda_1 \lambda_2 \lambda_3$$

$$I_1 = \lambda_1^2 + \lambda_2^2 + \lambda_3^2$$

and

$$I_2 = \lambda_1^2 \lambda_2^2 + \lambda_2^2 \lambda_3^2 + \lambda_3^2 \lambda_1^2$$

Thereby c_1, c_2 and D are material constants to be obtained from experiment. For incompressible Mooney–Rivlin solid, the Jacobian $J = \det \mathbf{F} = 1$ holds. The stress-strain relationship for Mooney–Rivlin material can be derived from the differentiation of the strain energy density function in Eq. (4.22) with respect to the deformation. Hence the Cauchy stress tensor takes the following form

$$\mathbf{S} = \frac{\partial \psi}{\partial \mathbf{B}} = -p\mathbf{I} + 2c_1 \mathbf{B} - 2c_2 \mathbf{B}^{-1} \qquad (4.24)$$

with the pressure p

$$p = \frac{2}{3}(c_1 I_1 - c_2 I_2)$$

For a case of incompressible Mooney–Rivlin material, a simple uniaxial loading produces the principal stretches, i.e. the stretch ratio $\lambda = l/l_0$, with $\lambda_1 = \lambda$, $\lambda_2 =$

$\lambda_3 = 1/\sqrt{\lambda}$. In terms of these principal stretches for uniaxial loading, the corresponding deformation gradient tensor $\mathbf{F}$ has the following form

$$(\mathbf{F}) = \begin{pmatrix} \lambda & 0 & 0 \\ 0 & 1/\sqrt{\lambda} & \\ 0 & 0 & 1/\sqrt{\lambda} \end{pmatrix}$$

Now the left Cauchy–Green deformation tensor $\mathbf{B} = \mathbf{F} \cdot \mathbf{F}^{\mathrm{T}}$ can be written as

$$(\mathbf{B}) = \begin{pmatrix} \lambda^2 & 0 & 0 \\ 0 & 1/\lambda & \\ 0 & 0 & 1/\lambda \end{pmatrix}$$

and the inverse of $\mathbf{B}$

$$(\mathbf{B}^{-1}) = \begin{pmatrix} 1/\lambda^2 & 0 & 0 \\ 0 & \lambda & \\ 0 & 0 & \lambda \end{pmatrix}$$

Substitution of these tensors and the pressure p

$$p = \frac{2}{3}[c_1(\lambda^2 + 2/\lambda) - c_2(2\lambda^2 + 1/\lambda^2)]$$

into Eq. (4.15) to get the Cauchy stress tensor yields

$$S = -p \begin{pmatrix} 1 & 0 & 0 \\ 0 & 1 & 0 \\ 0 & 0 & 1 \end{pmatrix} + 2c_1 \begin{pmatrix} \lambda^2 & 0 & 0 \\ 0 & 1/\lambda & 0 \\ 0 & 0 & 1/\lambda \end{pmatrix} - 2c_2 \begin{pmatrix} 1/\lambda^2 & 0 & 0 \\ 0 & \lambda & 0 \\ 0 & 0 & \lambda \end{pmatrix}$$

or

$$S = \begin{pmatrix} -p + 2c_1\lambda^2 - 2c_2/\lambda^2 & 0 & 0 \\ 0 & -p + 2c_1\lambda - 2c_2/\lambda & 0 \\ 0 & 0 & -p + 2c_1\lambda - 2c_2/\lambda \end{pmatrix}$$

Thereby

$$p = p(c_1, c_2; \lambda)$$

4.2.2 Ogden Materials

In the year 1072 R Ogden published a paper proposing a strain energy function ϕ for large deformation of isotropic and incompressible elastic materials.

$$\phi(\lambda_1, \lambda_2, \lambda_3; \alpha) = (\mu/\alpha)(\lambda_1^\alpha + \lambda_2^\alpha + \lambda_3^\alpha - 3)$$

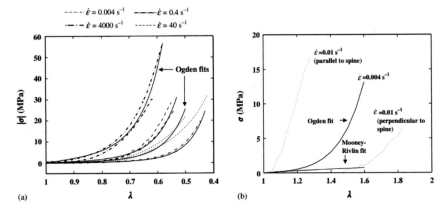

Fig. 4.2 Engineering stress vs stretch ratio of mammal skin to compression and tension (Shergold and Fleck 2006)

$\lambda_1, \lambda_2, \lambda_3$ are the principal stretches in all three directions, μ is initial shear modulus and α is a parameter to be varied empirically using related experiment. Ogden validated his theory using an experiment with rubber-like elastic solids.

With some adjustments the proposed strain energy function can be generalized as

$$\phi(\lambda_1, \lambda_2, \lambda_3; \alpha) = \sum_{i=1}^{N}(\mu/\alpha_i)(\lambda_1^{\alpha_i} + \lambda_2^{\alpha_i} + \lambda_3^{\alpha_i} - 3) + (K/2)(J - 1)^2$$

Here K is bulk modulus and J is the Jacobi determinant, which representing the relative volume $J = \lambda_1 \lambda_2 \lambda_3$.

Using the left Cauchy–Green tensor $\mathbf{B}$ the above strain energy function takes the following form

$$\psi(\mathbf{B}) = \sum_{i=1}^{N}(\mu_i/\alpha_i)\left((\mathbf{I} \cdot \cdot \mathbf{B})^{\alpha_i} - 3\right) + (K/2)(J - 1)^2$$

The Cauchy stress tensor $\mathbf{S}$ can be obtained from the gradient of ϕ with respect of the left Cauchy–Green tensor $\mathbf{B}$.

$$\mathbf{S} = \frac{\partial \psi}{\partial \mathbf{B}}$$

The material model proposed by Ogden proved to be suitable for modelling large deformation of isotropic elastic biomaterial as illustrated by Shergold and Fleck (2006). Figure 4.2 shows the response of mammal skin to large uniaxial compression and tension loading.

4.2.3 Superelastic Materials—Thamburaja and Nikabdullah Model

The term superelasticity is coined to refer to special elastic properties of smart material called shape memory alloy. Superelasticity or pseudoelasticity is load–deformation behaviour of shape memory alloy undergoing an isothermal process of phase transformation from austenite to martensite phase (forward transformation) and martensite to austenite phase (reverse transformation) occurred under an external load. If the forward transformation and reverse transformation routes do not coincide, i.e. if there is a hysteresis between them, then this behaviour is called pseudoelasticity. Figure 4.3 illustrates this behaviour. The bottom load-deformation diagram of Fig. 4.3 shows the shape memory effect of CuAlNi at low temperature, while the top diagram figure illustrates the complete superelastic characteristics of this material at high temperature.

There are several attempts to present a reliable model to capture this behaviour of shape memory alloy. In this section, we focus on the model proposed by Thamburaja and Nikabdullah (2009). The model is a constitutive model of shape memory alloy undergoing austenite-martensite phase transformation using the theory of isotropic metal plasticity.

The deformation of shape memory alloy, i.e. the martensitic transformation of its crystal structure, are subjected by two factors: mechanically induced transformation and temperature induced transformation. Therefore temperature plays important role in the superelasticity behaviour of shape memory alloy. The constitutive equation is then derived from the Helmholtz free energy to capture the influence of temperature in the deformation process. The total strain tensor, either left Cauchy–Green or right Cauchy-Green, is consisted of two parts, elastic and inelastic strain tensors, i.e.

$$\mathbf{B} = \mathbf{B}^e + \mathbf{B}^i$$

Now the Helmholz free energy can be written in separable, elastic and inelastic form, as

$$\psi = \psi(\mathbf{B^e}) + \psi^g + \psi^\xi + \psi^\theta$$

Where ψ^g the gradient free energy, responsible for a presence of austenite-martensite interfaces and ψ^ξ is the free energy associated with the martensitic transformation by amount of ξ martensite volume fraction and lastly free energy due to temperature induced martensitic transformation ψ^θ. In this section, we focus only on the Helmholz free energy resulted from mechanically induced martensitic transformation, $\psi(\mathbf{B}^e)$. This energy consisted of the elastic strain energy and the strain energy from thermal expansion of the material

$$\psi(\mathbf{B}^e) = \mu|\mathbf{B}_o^e|^2 + \kappa(\mathbf{I} \cdot \cdot \mathbf{B}^e)^2 - 3\kappa \, \alpha_{th}(\theta - \theta_o)(\mathbf{I} \cdot \cdot \mathbf{B}^e)$$

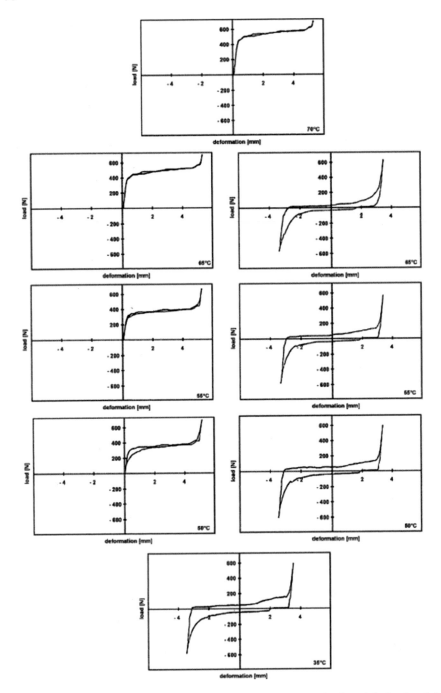

Fig. 4.3 Load–deformation diagram of superelastic material, CuAlNi Alloy, right heating, left cooling (Abdullah et al. 2002)

The constitutive equation derived from this free energy is the relation between the Cauchy stress tensor $\mathbf{S}$ and left Cauchy–Green tensor $\mathbf{B}^e$.

$$\mathbf{S} = \frac{\partial \psi}{\partial \mathbf{B}^e} = 2\mu \mathbf{B}^e_o + \kappa [\mathbf{I} \cdot \cdot \mathbf{B}^e - 3\alpha_{th}(\theta - \theta_o)]\mathbf{I}$$

$\mathbf{B}^e_o$ and θ_o are the initial strain tensor and initial temperature, respectively.

4.2.4 Examples and Problems

Example 4.2.1

Verify the relationship between the Lamè constants and the experimentally determined the elastic and bulk moduli, E, G and the Poisson ratio v as described by Eq. (4.13).

Solution

We can start with the Lame–Navier equation (4.12) and use it for a simple one-dimensional extension test. That is

$$\sigma_{pq} = \lambda \delta_{pq} \epsilon_{ss} + 2\mu(\epsilon_{pq} + \epsilon_{qp}) \longleftrightarrow \mathbf{S} = \lambda \, tr(\hat{\mathbf{E}})\mathbf{I} + 2\mu \hat{\mathbf{E}}$$

and the Cauchy stress tensor for a simple one-dimensional extension is given as follows:

$$(\mathbf{S}) = \begin{pmatrix} \sigma_{11} & 0 & 0 \\ 0 & 0 & 0 \\ 0 & 0 & 0 \end{pmatrix}$$

The linearized Lagrangian strain tensor for extension test takes the following form

$$(\hat{\mathbf{E}}) = \begin{pmatrix} \epsilon_{11} & 0 & 0 \\ 0 & \epsilon_{22} & 0 \\ 0 & 0 & \epsilon_{33} \end{pmatrix}$$

Let $\sigma_{11} = P$ the stress along the extension axis, then $\epsilon_{11} = \frac{P}{E}$ and $\epsilon_{22} = \epsilon_{33} = -\frac{vP}{E}$. Where E and v are Young's modulus and Poisson ratio, respectively. Therefore, the strain tensor now is

$$(\hat{\mathbf{E}}) = \begin{pmatrix} \frac{P}{E} & 0 & 0 \\ 0 & -\frac{vP}{E} & 0 \\ 0 & 0 & -\frac{vP}{E} \end{pmatrix}$$

Substituting these two tensors into the Lame–Navier Equation (4.12), we have

$$\begin{pmatrix} P\,0\,0 \\ 0\,0\,0 \\ 0\,0\,0 \end{pmatrix} = \lambda(\frac{P}{E} - 2\frac{vP}{E})\begin{pmatrix} 1\,0\,0 \\ 0\,1\,0 \\ 0\,0\,1 \end{pmatrix} + 2\mu\begin{pmatrix} \frac{P}{E} & 0 & 0 \\ 0 & -\frac{vP}{E} & 0 \\ 0 & 0 & -\frac{vP}{E} \end{pmatrix}$$

and after simplification

$$P = \lambda(\frac{P}{E} - 2\frac{vP}{E}) + 2\frac{vP}{E}$$

and

$$0 = \lambda(\frac{P}{E} - 2\frac{vP}{E}) - 2\mu\frac{vP}{E}$$

By eliminating P, we obtain the relationship between the Lame constants and the elasticity constant E and v.

$$\lambda = \frac{vE}{(1 - v)(1 - 2v)}$$

$$\mu = \frac{E}{2(1 + v)} = G$$

Example 4.2.2

The deformation gradient of a simple shear test is given as follows:

$$(\mathbf{F}) = \begin{pmatrix} 1\,\gamma\,0 \\ 0\,1\,0 \\ 0\,0\,1 \end{pmatrix}$$

γ is the shear strain and in term of stretch ratio $\lambda = l/l_0$, it takes the following form

$$\gamma = \lambda - 1/\lambda$$

furthermore

$$\lambda_1 = \lambda, \lambda_2 = 1/\lambda, \lambda_3 = 1$$

Plot the Cauchy stress tensor versus stretch ratio if the sample for the testing used is a rubber block modelled as an incompressible Mooney–Rivlin material with $c_{10} = c_1 = 0.220 MPa$ and $c_{01} = c_2 = 0.017 MPa$. Determine the Cauchy stress if the sample is stretched to 30 % of its initial length.

Solution

From the given deformation gradient of a simple shear

$$(\mathbf{F}) = \begin{pmatrix} 1 & \gamma & 0 \\ 0 & 1 & 0 \\ 0 & 0 & 1 \end{pmatrix}$$

We determine firstly the right Cauchy–Green deformation tensor $\mathbf{B}$, together with its inverse $\mathbf{B}^{-1}$ to get the Cauchy stress tensor according to Eq. (4.15). After some linear algebraic manipulations, we have the corresponding right Cauchy–Green deformation tensor as follows:

$$(\mathbf{B}) = (\mathbf{F} \cdot \mathbf{F}^{\mathsf{T}}) = \begin{pmatrix} 1+\gamma^2 & \gamma & 0 \\ \gamma & 1 & 0 \\ 0 & 0 & 1 \end{pmatrix}$$

and its inverse $\mathbf{B}^{-1}$

$$(\mathbf{B}^{-1}) = \begin{pmatrix} 1 & -\gamma & 0 \\ -\gamma & 1+\gamma^2 & 0 \\ 0 & 0 & 1 \end{pmatrix}$$

Putting all these tensors together in Eq. (4.15), we have the Cauchy stress tensor

$$(\mathbf{S}) = \begin{pmatrix} -p + (c_1 - c_2) + 2c_2\gamma^2 & 2(c_1 + c_2)\gamma & 0 \\ 2(c_1 + c_2)\gamma & -p + (c_1 - c_2) - 2c_2\gamma^2 & 0 \\ 0 & 0 & -p + (c_1 - c_2) \end{pmatrix}$$

and the pressure p is given as

$$p = \frac{2}{3}(c_1 I_1 - c_2 I_2)$$

$$= \frac{2}{3}[c_1(3 + \gamma^2) - c_2(3 + \gamma^2)]$$

$$= \frac{2}{3}(c_1 - c_2)(3 + \gamma^2)$$

With $c_1 = 0.220 M Pa$ and $c_2 = 0.017 M Pa$ the normal stress, for instance σ_{11} is

$$\sigma_{11} = -\frac{2}{3}(c_1 - c_2)(3 + \gamma^2) + (c_1 - c_2) + 2c_2\gamma^2$$

After simplification and substitution, we have

$$\sigma_{11} = -0.203 - 0.101\gamma^2$$

Since γ^2 is always positive, the normal stress for simple shear is always negative, thus indicating compression. For the case of stretch ratio equal to 30%, $\gamma = 30 - 1/30 = 29.97$. Therefore the compression stress is $90.91 MPa$.

Problem 4.2.1

An uniaxial tension test has been carried out to a sample of rubber with varying stretch ratio λ up to 2.5. Rubber sample is assumed to behave as a Mooney–Rivlin material with the strain energy function

$$\psi = c_1(I_{\mathbf{B}} - 3) + c_2(II_{\mathbf{B}} - 3)$$

and the Cauchy stress tensor

$$\mathbf{S} = \frac{\partial \psi}{\partial \mathbf{B}} = -p\mathbf{I} + 2c_1\mathbf{B} - 2c_2\mathbf{B}^{-1}$$

The material parameters are given as $c_{10} = c_1 = 0.916MPa$ and $c_{01} = c_2 = 0.0647MPa$. Plot a graph of normal stress versus stretch ratio in the direction (1).

Problem 4.2.2

In case of two-dimensional plane state of stress the right Cauchy–Green strain tensor is given as follows (Fig. 4.4):

$$(\mathbf{C}) = \begin{pmatrix} c_{12} & c_{12} & 0 \\ c_{21} & c_{22} & 0 \\ 0 & 0 & c_{33} \end{pmatrix}$$

Thereby

$$c_{33} = \frac{h^2}{H^2}$$

and the constants h and H are the current and reference thicknesses, respectively. Show that the pressure of an incompressible neo-Hookean material (i.e. Mooney–Rivlin material for plane state of stress omitting the material constant c_2 and $S_{33} = 0$) can be expressed as follows:

$$p = \frac{1}{3}\mu(I_1 - 3c_{33})$$

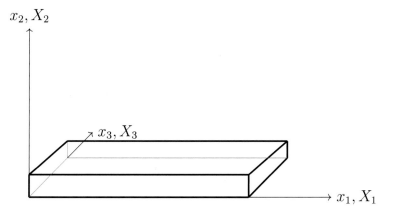

Fig. 4.4 Uniaxial tension test

Problem 4.2.3

Experimental data of a isotropic incompressible rubber-like animal skin material are sometimes correlated by an Ogden model. The strain energy function of an Odgen material is defined as follows:

$$\psi(\lambda_1, \lambda_2, \lambda_3) = \sum_{i=1}^{n} \frac{\mu_i}{\alpha_i} (\lambda_1^{\alpha_i} + \lambda_2^{\alpha_i} + \lambda_3^{\alpha_i} - 3)$$

Problem 4.2.4

A modified version of St. Venant–Kirchhoff constitutive behaviour is defined by the following strain energy density function ψ as a scalar function of the Green–Lagrange strain tensor $\mathbf{E}$ and the Jacobian of the deformation gradient J

$$\Psi(\mathbf{E}, J) = \mu\, tr(\mathbf{E}^2) + \frac{\kappa}{2}(\ln J)^2$$

μ and κ are material constants.

1. Obtain an expression for second Piola–Kirchhoff stress tensor $\hat{\mathbf{T}}$ as a function of the right Cauchy–Green tensor $\mathbf{C}$.
2. Obtain an expression for first Piola–Kirchhoff stress tensor $\mathbf{T}$ as a function of the left Cauchy–Green tensor $\mathbf{B}$.
3. Determine the material elasticity tensor.

Problem 4.2.5

Beginning from the generalized Hooke's law

$$S(x, t) = \underline{E} : \hat{E}(x, t) \longleftrightarrow \sigma_{ij}(x_i, t) = E_{ijkl}\epsilon_{kl}(x_i, t) \tag{4.25}$$

derive the associated strain energy function ψ.

Problem 4.2.6

If γ is the shear strain measured in a simple two-dimensional shear test of an linear-elastic body. Show that the principal stresses in one particular point in a body can be expressed as

$$\sigma_1 = -\sigma_2 = 2G \sinh^{-1}\frac{\gamma}{2}$$

G is the bulk modulus of the material.

Problem 4.2.7

A rubber block is subjected to a simple shear test with the deformation gradient F

$$F = \begin{pmatrix} 1 & \gamma & 0 \\ 0 & 1 & 0 \\ 0 & 0 & 1 \end{pmatrix}$$

γ is the shear strain. The material property of the rubber is modelled by a two-parameter Mooney–Rivlin hyperelastic material model with the strain energy density given as follows:

$$W = c_{10}(\hat{I}_1 3) + c_{01}(\hat{I}_2 - 3) + (1/D_1)(J - 1)^2$$

J is the Jacobian parameter. Taking the following data into consideration calculate the resulting 2nd Piola–Kirchhoff stress tensor.

$$c_{10} = -0.55 MPa, c_{01} = 0.7 MPa, D_1 = 0.001(MPa)^{-1}$$

Problem 4.2.8

A thin sheet of incompressible hyperelastic material is subjected to a biaxial tension test with the state of plane stress, i.e. $\hat{t}_{23} = \hat{t}_{13} = \hat{t}_{33} = 0$, show that

$$\hat{t}_{11} = 2\left(\frac{\partial \psi}{\partial I_C} + (\lambda_1^2 + \lambda_2^2 + \frac{1}{\lambda_1^2 \lambda_2^2})\frac{\partial \psi}{\partial I I_C}\right) - \lambda_1^2 \frac{\partial \psi}{\partial I I_C}$$

$$\hat{t}_{22} = 2\left(\frac{\partial \psi}{\partial I_C} + (\lambda_1^2 + \lambda_2^2 + \frac{1}{\lambda_1^2 \lambda_2^2})\frac{\partial \psi}{\partial I I_C}\right) - \lambda_2^2 \frac{\partial \psi}{\partial I I_C}$$

Where $\hat{t}_{11} = \hat{t}_1$ and $\hat{t}_{22} = \hat{t}_2$ are the elements of second Piola–Kirchhoff stress in the principal axes and the I_C and II_C are the invariants of the right Cauchy–Green deformation tensor $\mathbf{C}$. The right and left stretch tensor of a biaxial tension is given as follows:

$$\mathbf{F} = \mathbf{U} = \mathbf{V} = \begin{pmatrix} \lambda_1 & 0 & 0 \\ 0 & \lambda_2 & 0 \\ 0 & 0 & \frac{1}{\lambda_1 \lambda_2} \end{pmatrix}$$

Chapter 5
Balance Principles in Continuum Mechanics

The mechanical balance or conservation principles in continuum mechanics deal with auditing the transfer of specific mechanical properties from a system to its environment or vice versa through the system boundary. The mechanical properties referred to here are the mass, linear momentum, angular momentum and energy.

5.1 General Mechanical Balance Equation

Let $\mathcal{A}$ be an arbitrary property of a material to be transported and $\psi(\mathbf{x}, t)$ be the amount of the property per unit of mass. The spatial control volume is denoted by $\mathcal{V}$ and its surface by $\partial\mathcal{V}$. The rate of change per unit of time t of the amount of the property $\mathcal{A}$ in the control volume $\mathcal{V}$ is governed by three factors. They are

- generation of the property per unit mass and time by a source $p_A(\mathbf{x}, t)$ inside the control volume,
- the convective flux (net incoming) of the amount of property across the surface $\partial\mathcal{V}$ of the control volume and
- the non-convective (net incoming) flux across the source of the control volume $\mathbf{j}_A(\mathbf{x}, t)$ such as radiation flux, magnetic and electrical flux.

The differential amount of property transferred convectively across the boundary of the control volume is given as follows:

$$d\Psi_{\partial\mathcal{V}} = \frac{\psi \, dm}{dt} = \rho\psi (\mathbf{v} \cdot \mathbf{n}) \, dA$$

while the non-convective term, i.e. transfer of property without any involvement of material motion, takes the following form:

Fig. 5.1 Transfer of
property

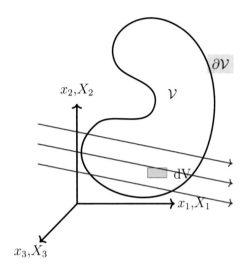

$$dJ_A = j_A \cdot n \, dA$$

Thereby ρ is the density of the material. The general balance equation in global form for the transport of the property $\psi(\mathbf{x}, t)$ across a system boundary can be as expressed as follows:

$$\frac{\partial}{\partial t}\left(\int_V \rho \, \psi \, dV\right) = \int_V \rho \, p_A dV - \int_{\partial V} \rho \, \psi(\mathbf{v} \cdot \mathbf{n}) \, dA - \int_V j_A \cdot \mathbf{n} \, dA \quad (5.1)$$

For the mechanical balance equation, Eq. (5.1) is reduced to

$$\frac{\partial}{\partial t}\left(\int_V \rho \, \psi \, dV\right) = \int_V \rho \, p_A dV - \int_{\partial V} \rho \, \psi(\mathbf{v} \cdot \mathbf{n}) \, dA \quad (5.2)$$

without non-convective components, such as heat transfer via conduction, electrical flow flux and magnetic flux (Fig. 5.1).

5.1.1 Balance Equation for Mass Transfer

For the case the property $\mathcal{A}$ is a mass, the amount of mass per unit mass is unity, i.e. $\psi \equiv 1$ then the mechanical balance equation (5.2) becomes

$$\frac{\partial}{\partial t}\left(\int_V \rho \, dV\right) = \int_V \rho \, p_A dV - \int_{\partial V} \rho \, (\mathbf{v} \cdot \mathbf{n}) \, dA$$

It is assumed that the control volume is time-independent and no generation of mass in the control volume, i.e. $p_A = 0$. By applying the divergence theorem, the above equation can be simplified further to a local form of mass transfer equation

$$\left(\int_V \frac{\partial \rho}{\partial t} dV \right) = \int_V \rho \, p_A dV - \int_{\partial V} \nabla \cdot (\rho \mathbf{v}) \, dV$$

$$\frac{\partial \rho}{\partial t} + \nabla \cdot (\rho \mathbf{v}) = 0 \tag{5.3}$$

Equation (5.3) is well known under the name the continuity equation. It is a local form of mass transfer balance equation.

5.1.2 Balance Equation for Linear Momentum

If the property $\mathcal{A}$ is a linear momentum, then the amount of linear momentum per unit mass is the velocity, i.e. $\psi \equiv \mathbf{v}$. Therefore, the left-hand side (LHS) of the equations (5.1) or (5.2) becomes an expression of forces:

$$\frac{\partial}{\partial t} \left(\int_V \rho \mathbf{v} \, dV \right) = \mathbf{f}$$

The force $\mathbf{f}$ resulting from the rate of change of linear momentum is balanced by the resultant of the body and surface forces in and on the body (control volume):

$$\mathbf{f} = \int_V \rho \, \mathbf{b} dV + \int_{\partial V} \rho \, \mathbf{t} \, dA$$

where $\mathbf{b}$ and $\mathbf{t}$ are the body force and traction or stress vector, respectively. Now with the Cauchy stress tensor $\mathbf{S}$, we can express the traction vector as

$$\mathbf{t} = \mathbf{n} \cdot \mathbf{S}^T = \mathbf{n} \cdot \mathbf{S}$$

Combining the above equation together and applying the divergence theorem, we have the balance equation for linear momentum as follows:

$$\frac{\partial}{\partial t} \left(\int_V \rho \mathbf{v} \, dV \right) = \int_V \rho \, \mathbf{b} dV + \int_{\partial V} \mathbf{n} \cdot \mathbf{S} \, dA$$

$$\frac{\partial}{\partial t} \left(\int_V \rho \mathbf{v} \, dV \right) = \int_V \rho \, \mathbf{b} dV + \int_V \nabla \cdot \mathbf{S} \, dV \tag{5.4}$$

The corresponding local form of Eq. (5.4) is as follows:

$$\frac{\partial(\rho\mathbf{v})}{\partial t} = \nabla \cdot \mathbf{S} + \rho\mathbf{b} \tag{5.5}$$

5.1.3 Balance Equation for Angular Momentum

For the rate of change of angular momentum, the amount of property per unit mass is given as $\psi \equiv \mathbf{r} \times \mathbf{v}$, so that the LHS of the equation (5.1)

$$\frac{\partial}{\partial t}\left(\int_{\mathcal{V}} \rho\mathbf{r} \times \mathbf{v}\, dV\right) = \mathcal{M}$$

This moment $\mathcal{M}$ is balanced by the moments produced by the body and traction forces of the body:

$$\mathcal{M} = \int_{\mathcal{V}} \rho\mathbf{r} \times \mathbf{b}\, dV + \int_{\partial\mathcal{V}} \mathbf{r} \times \mathbf{t}\, dA$$

$$= \int_{\mathcal{V}} \rho\mathbf{r} \times \mathbf{b}\, dV + \int_{\partial\mathcal{V}} \mathbf{r} \times (\mathbf{S}^{\mathbf{T}} \cdot \mathbf{n})\, dA$$

After applying the divergence theorem

$$\mathcal{M} = \int_{\mathcal{V}} (\rho\mathbf{r} \times \mathbf{b}\, dV + (\mathbf{r} \times \mathbf{S}^{\mathbf{T}}) \cdot \nabla)dV$$

with

$$(\mathbf{r} \times \mathbf{S}^{\mathbf{T}}) \cdot \nabla = (\mathbf{r} \times \nabla) \cdot \mathbf{S}^{\mathbf{T}} + \mathbf{r} \times (\mathbf{S}^{\mathbf{T}} \cdot \nabla) = (\mathbf{r} \times \nabla) \cdot \mathbf{S}^{\mathbf{T}} + \mathbf{r} \times (\nabla \cdot \mathbf{S})$$

Inserting this identity into the equation for the moment due to body and traction forces, we have

$$\mathcal{M} = \int_{\mathcal{V}} (\rho\mathbf{r} \times \mathbf{b} + \mathbf{r} \times (\nabla \cdot \mathbf{S}) + (\mathbf{r} \times \nabla) \cdot \mathbf{S}^{\mathbf{T}})dV$$

which reduces to only

$$0 = \int_{\mathcal{V}} (\mathbf{r} \times \nabla) \cdot \mathbf{S}^{\mathbf{T}} dV \tag{5.6}$$

due to

$$\int_{\mathcal{V}} (\rho\mathbf{r} \times \mathbf{b} + \mathbf{r} \times (\nabla \cdot \mathbf{S})\, dV = \frac{\partial}{\partial t}\left(\int_{\mathcal{V}} \rho\mathbf{r} \times \mathbf{v}\, dV\right), \quad \forall\mathbf{r}$$

as stated in Eq. (5.4), the balance equation for linear momentum. Equation (5.6) is only valid if

$$(\mathbf{r} \times \nabla) \cdot \mathbf{S}^T = \mathbf{0}$$

which implies the symmetry of the Cauchy stress tensor, i.e.

$$\mathbf{S} = \mathbf{S}^T \tag{5.7}$$

Equation (5.7) above is also known as Cauchy 2 while the local form of the balance equation of linear momentum described by Eq. (5.5) is called Cauchy 1.

5.1.4 Examples and Problems

Problem 5.1.1

Given is the Cauchy stress tensor $\mathbf{S}$ as follows:

$$(\mathbf{S}) = \begin{pmatrix} x_1 + x_2 & f(x_1, x_2) & 0 \\ f(x_1, x_2) & x_1 - 2x_2 & 0 \\ 0 & 0 & x_2 \end{pmatrix}$$

1. Determine the function $f(x_1, x_2)$ assuming that the body force is negligibly small and the origin is stress-free.
2. Determine the stress vector at any point on the surface $x_1 = constant$ and $x_2 = constant$.

Problem 5.1.2

If $\mathbf{T}$ is any symmetrical tensor field, i.e. $\mathbf{T} = \mathbf{T}(x_i)$, $i = 1, 2, 3$, is then $\mathbf{S}$, with

$$\mathbf{S} = \nabla \times (\nabla \times \mathbf{T})$$

statically allowable to represent the Cauchy stress tensor.

Problem 5.1.3

Verify the relation

$$(\mathbf{r} \times \mathbf{S}^T) \cdot \nabla = (\mathbf{r} \times \nabla) \cdot \mathbf{S}^T + \mathbf{r} \times (\mathbf{S}^T \cdot \nabla) = (\mathbf{r} \times \nabla) \cdot \mathbf{S}^T + \mathbf{r} \times (\nabla \cdot \mathbf{S})$$

Problem 5.1.4

Show that

$$(\mathbf{r} \times \nabla) \cdot \mathbf{S}^\mathrm{T} = \mathbf{0} \longleftrightarrow \mathbf{S} = \mathbf{S}^\mathrm{T}$$

Problem 5.1.5

Show that the Cauchy stress tensor possesses only real eigenvalues.

Problem 5.1.6

Write down an equation of motion either symbolically or judicially for a moving material whose state of stress is everywhere a pure hydrostatic one.

Problem 5.1.7

A motion is referred to as potential motion if its velocity $\mathbf{v}$ can be expressed as

$$\mathbf{v} = \nabla \phi$$

with ϕ a scalar function of position.

Derive the continuity equation for such motion of a compressible and an incompressible medium in terms of the potential ϕ.

Problem 5.1.8

Given is the velocity field of a two-dimensional flow as follows:

$$\mathbf{v}(x, y) = (ax + b)\mathbf{e_x} + (-ay + c)\mathbf{e_y}$$

Find the potential function ϕ of the flow and verify that it obeys the Laplace equation.

Problem 5.1.9

Show that the planar state of stress can be described by the following equation:

$$\sigma_{11} = \psi_{,22} + \kappa, \ \sigma_{22} = \psi_{,11} + \kappa, \ \sigma_{12} = -\psi_{,12}$$

whereby $\psi(x_1, x_2)$ is the stress function, also known as Airy's stress function and κ is a potential field.

Chapter 6
Solutions to the Problems

6.1 Section 2.1

Example 2.1.1

Examine if the following relation between the Lagrangian and Eulerian coordinates represents the motion.

$$x^1 = X^1 e^{(t)} + X^3 (e^{(t)} - 1), \quad x^2 = X^2 + X^3 (e^{(t)} - e^{(-t)}), \quad x^3 = X^3$$

Solution

The solution steps are (i) finding the deformation gradient $\mathbf{F} = \frac{\partial x^k}{\partial X^l}$ and (ii) checking if the Jacobi determinant J is positive definite.

From the relation above, we have

$$\frac{\partial x^1}{\partial X^1} = e^{(t)}; \quad \frac{\partial x^1}{\partial X^2} = 0; \quad \frac{\partial x^1}{\partial X^3} = e^{(t)} - 1$$

etc.

Altogether, we have then the matrix of the deformation gradient

$$\mathbf{F} = \frac{\partial x^p}{\partial X^q} = \begin{pmatrix} e^{(t)} & 0 & e^{(t)} - 1 \\ 0 & 1 & (e^{(t)} - e^{(-t)}) \\ 0 & 0 & 1 \end{pmatrix}$$

Therefore, the Jacobi determinant $J = det((\mathbf{F}))$

© The Author(s), under exclusive license to Springer Nature Singapore Pte Ltd. 2023
N. A. N. Mohamed, *Introduction to Continuum Mechanics for Engineers*,
https://doi.org/10.1007/978-981-99-0811-0_6

$$J = \det \begin{pmatrix} e^{(t)} & 0 & e^{(t)} - 1 \\ 0 & 1 & (e^{(t)} - e^{(-t)}) \\ 0 & 0 & 1 \end{pmatrix} = e^{(t)}$$

which is always positive for all time t. So the relation given above represents a motion.

Example 2.1.2

Given is the temperature field described using the spatial, Eulerian coordinates as follows:

$$T = T_0(x^1 + x^2)$$

The plane motion of the continuum is characterized by the relation

$$x^1 = X^1 + At X^2, x^2 = X^2, x^3 = X^3$$

T_0 and A are given constants.

1. Express this temperature field in terms of Lagrangian coordinates.
2. Determine the material derivative of this temperature field in both coordinate systems.

Solution

(a) A direct substitution of Lagrangian coordinates into the equation for the given temperature field yields the following equation:

$$T = T_0(X^1 + At X^2 + X^2) = T_0(X^1 + (At + 1)X^2)$$

(b) Material derivative in Lagrangian formalism

$$\frac{DT}{Dt} = \frac{\partial T(X^k, t)}{\partial t}\Big|_{X^k} = AT_0X^2 = AT_0x^2$$

Material derivative in Eulerian formalism

$$\frac{DT}{Dt} = \frac{\partial T(x^k, t)}{\partial t}\Big|_{x^k} + \frac{\partial T(x^i, t)}{\partial x^k}\frac{dx^k}{dt}$$

From the given plane motion, we have

$$\frac{dx^1}{dt} = AT_0 X^2 = AT_0 x^2$$

$$\frac{dx^2}{dt} = 0$$

and

$$\frac{dx^3}{dt} = 0$$

$$\frac{\partial T(x^k, t)}{\partial t}\Big|_{x^k} = 0$$

and furthermore

$$\left(\frac{\partial T(x^i, t)}{\partial x^k}\right) = \begin{pmatrix} T_0 \\ T_0 \\ 0 \end{pmatrix}$$

Thus, the material derivative in Eulerian formalism is

$$\frac{DT}{Dt} = 0 + (Ax^2, 0, 0) \begin{pmatrix} T_0 \\ T_0 \\ 0 \end{pmatrix} = AT_0 x^2$$

Both calculations lead to the same result however in the Lagrangian formalism, it is much simpler and straightforward.

Problem 2.1.1

Given is a scalar field $\phi = \phi_0(x^1 - 2x^2)$ represented by spatial, Eulerian coordinates. The motion of the continuum associated with the above field is given as follows:

$$x^1 = X^1 + t^2 X^2, x^2 = -t^2 X^1 + X^2, x^3 = X^3$$

Find the material derivative of the scalar field ϕ in both Lagrangian and Eulerian formalisms. Compare the results.

Solution

(a) A direct substitution of Lagrangian coordinates into the equation for the given scalar field ϕ yields the following equation:

$$\phi = \phi_0((X^1 + t^2 X^2) - 2(-t^2 X^1 + X^2)) = \phi_0((1 + t^2)X^1 + (t^2 - 2)X^2)$$

(b) Material derivative in Lagrangian formalism

$$\frac{D\phi}{Dt} = \frac{\partial\phi(X^k, t)}{\partial t}|_{X^k} = 2t\phi_0(2X^1 + X^2)$$

Material derivative in Eulerian formalism

$$\frac{D\phi}{Dt} = \frac{\partial\phi(x^k, t)}{\partial t}|_{x^k} + \frac{\partial\phi(x^i, t)}{\partial x^k}\frac{dx^k}{dt}$$

From the given motion, we have

$$\frac{dx^1}{dt} = 2t X^2$$

$$\frac{dx^2}{dt} = -2t X^1$$

and

$$\frac{dx^3}{dt} = 0$$

$$\frac{\partial\phi(x^k, t)}{\partial t}|_{x^k} = 0$$

and furthermore

$$(\frac{\partial\phi(x^i, t)}{\partial x^k}) = \begin{pmatrix} 2t X^2 \\ -2t X^1 \\ 0 \end{pmatrix}$$

Thus, the material derivative in Eulerian formalism is

$$\frac{D\phi}{Dt} = 0 + (\phi_0, -2\phi_0, 0)\begin{pmatrix} 2t X^2 \\ -2t X^1 \\ 0 \end{pmatrix} = 2t\phi_0(2X^1 + X^2)$$

Both formalisms produce the same results.

Problem 2.1.2

One particular motion is characterized by the following relation between material and spatial coordinates

$$x^1 = X^1 + \alpha X^2, x^2 = X^2, x^3 = X^3$$

Determine the associated deformation gradient, the Jacobian, the metric tensors and Lagrangian strain tensor. Discuss the special cases of $\alpha = 0$ and $\alpha = 1$.

Solution

The element of the deformation gradient $\mathbf{F}$

$$(\mathbf{F}) = \begin{pmatrix} 1 & \alpha & 0 \\ 0 & 1 & 0 \\ 0 & 0 & 1 \end{pmatrix}$$

The Jacobian J is calculated from $det((\mathbf{F}))$

$$J = det \begin{pmatrix} 1 & \alpha & 0 \\ 0 & 1 & 0 \\ 0 & 0 & 1 \end{pmatrix} = 1$$

The matrix tensors, i.e. the right Cauchy–Green tensor $\mathbf{F}^T \cdot \mathbf{F} = \mathbf{C}$ and the left Cauchy–Green tensor $\mathbf{F} \cdot \mathbf{F}^T = \mathbf{B}$ for this motion are

$$(\mathbf{C}) = (\mathbf{F}^T \cdot \mathbf{F}) = \begin{pmatrix} 1 & \alpha & 0 \\ \alpha & \alpha^2 + 1 & 0 \\ 0 & 0 & 1 \end{pmatrix}$$

$$(\mathbf{B}) = (\mathbf{F} \cdot \mathbf{F}^T) = \begin{pmatrix} \alpha^2 + 1 & \alpha & 0 \\ \alpha & 1 & 0 \\ 0 & 0 & 1 \end{pmatrix}$$

The Lagrangian strain tensor $\mathbf{E}$ is calculated from $2\,\mathbf{E} = \mathbf{C} - \mathbf{I}$

$$(\mathbf{E}) = \begin{pmatrix} \alpha^2 & \alpha & 0 \\ \alpha & 0 & 0 \\ 0 & 0 & 0 \end{pmatrix}$$

For the case of $\alpha = 0$, there is rigid body motion without any deformation is called shearing or shear for the case of $\alpha = 1$.

6.2 Section 2.2

Example 2.2.1

St. Venant torsion is described by the small rotation of the beam's cross-sectional area and the warping of that area in the axial direction. The warping of the cross-sectional area is independent of axial axis X_3 but completely a function of cross-sectional coordinates X_1, X_2 or r, $\theta(X_3)$, with $\theta(X_3)$ is the cross-sectional twist angle per length of the beam. Based on reference configuration the "motion", St. Venant torsion is described by the following relation:

$$x_1 = X_1 - \frac{\theta}{l}X_2X_3$$

$$x_2 = X_2 + \frac{\theta}{l}X_3X_1$$

$$x_3 = X_3 + \psi(X_1, X_2)$$

whereby l is the length of the beam and $\psi(X_1, X_2)$ is the warping function representing the warping of the beam's cross section. Find the displacement vector $\mathbf{u}$. Determine the deformation gradient $\mathbf{F}$ and the corresponding Lagrangian displacement gradient $\mathbf{H}$. Determine also the Eulerian displacement gradient $\overset{\ast}{\mathbf{H}}$ (Fig. 6.1).

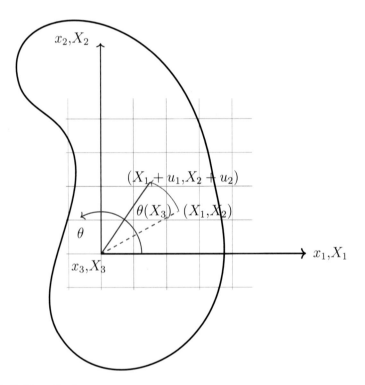

Fig. 6.1 St. Venant torsion

Solution

From the equations relating the St. Venant torsion in both material and spatial coordinates, X_i and x_j, $i, i = 1, 2, 3$

$$x_1 = X_1 - \frac{\theta}{l} X_2 X_3$$

$$x_2 = X_2 + \frac{\theta}{l} X_3 X_1$$

$$x_3 = X_3 + \psi(X_1, X_2)$$

The displacement vector u_i can be obtained directly using the equation

$$u_i = x_i - X_i$$

which produces

$$u_1 = -\frac{\theta}{l} X_2 X_3$$

$$u_2 = \frac{\theta}{l} X_3 X_1$$

$$u_3 = \psi(X_1, X_2)$$

the components of the displacement vector.

We have the components of the deformation gradient, $\partial x_i / \partial X_j$ by differentiating x_i with respect to material coordinates X_j, as follows:

$$\partial x_1 / \partial X_1 = 1$$

$$\partial x_1 / \partial X_2 = -\frac{\theta}{l} X_3$$

$$\partial x_1 / \partial X_2 = -\frac{\theta}{l} X_2$$

$$\partial x_2 / \partial X_1 = \frac{\theta}{l} X_3$$

$$\partial x_2 / \partial X_2 = 1$$

$$\partial x_2 / \partial X_3 = \frac{\theta}{l} X_1$$

$$\partial x_3 / \partial X_1 = \partial \psi(X_1, X_2) / \partial X_1$$

$$\partial x_3 / \partial X_2 = \partial \psi(X_1, X_2) / \partial X_2$$

$$\partial x_3 / \partial X_3 = 1$$

Rearranging the equations above, we have then the deformation gradient $\mathbf{F}$

$$(\mathbf{F}) = \partial x_i/\partial X_j = \begin{pmatrix} 1 & -(\theta/l)X_3 & -(\theta/l)X_2 \\ (\theta/l)X_3 & 1 & (\theta/l)X_1 \\ \partial\psi(X_1, X_2)/\partial X_1 & \partial\psi(X_1, X_2)/\partial X_2 & 1 \end{pmatrix}$$

Now to find the displacement gradient, we use Eq. (2.7) and get

$$(\mathbf{H}) = \partial u_i/\partial X_j = \begin{pmatrix} 0 & -(\theta/l)X_3 & -(\theta/l)X_2 \\ (\theta/l)X_3 & 0 & (\theta/l)X_1 \\ \partial\psi(X_1, X_2)/\partial X_1 & \partial\psi(X_1, X_2)/\partial X_2 & 0 \end{pmatrix}$$

Of course, we can also obtain the displacement gradient from the calculated displacement vector $\mathbf{u}$.

Problem 2.2.1

Given is the following motion:

$$x_1 = X_1$$
$$x_2 = X_1 + X_2$$
$$x_3 = X_1 + X_2 + X_3$$

Verify that the above equations represent a motion. Calculate the displacement gradient in both Lagrangian and Eulerian approaches.

Solution

The deformation gradient $\mathbf{F}$ is obtained from the above equations:

$$(\mathbf{F}) = \partial x_i/\partial X_j = \begin{pmatrix} 1 & 0 & 0 \\ 1 & 1 & 0 \\ 1 & 1 & 1 \end{pmatrix}$$

The Jacobian J of the deformation gradient above is equal to 1 which suggests that the above relation represents a motion. The inverse of the deformation gradient, $\mathbf{F}^{-1}$, is calculated as follows:

$$(\mathbf{F}^{-1}) = \partial X_i/\partial x_j = \begin{pmatrix} -1 & 1 & 0 \\ 1 & 0 & 0 \\ 0 & -1 & 0 \end{pmatrix}$$

The Lagrangian displacement tensor **H**, $\mathbf{H} = \mathbf{F} - \mathbf{I}$

$$(\mathbf{H}) = (\mathbf{F} - \mathbf{I}) = \begin{pmatrix} 1 & 0 & 0 \\ 1 & 0 & 0 \\ 1 & 1 & 0 \end{pmatrix}$$

and the Eulerian displacement tensor $\hat{\mathbf{H}}$, $\hat{\mathbf{H}} = \mathbf{I} - \mathbf{F}^{-1}$

$$(\hat{\mathbf{H}}) = (\mathbf{I} - \mathbf{F}^{-1}) = \begin{pmatrix} 2 & -1 & 0 \\ -1 & 1 & 0 \\ -1 & 2 & 0 \end{pmatrix}$$

Problem 2.2.2

Given is a body with the motion

$$x_1 = 1 + (X_1)^2 X_2$$
$$x_2 = 1 + (X_2)^2 X_1$$
$$x_3 = X_3$$

Perform the task as in the Problem 2.2.1. Verify that the above equations represent a motion. Calculate the displacement gradient in both Lagrangian and Eulerian approaches.

Solution

The deformation gradient **F**

$$(\mathbf{F}) = \partial x_i / \partial X_j = \begin{pmatrix} 2X_1X_2 & (X_1)^2 & 0 \\ (X_2)^2 & 2X_1X_2 & 0 \\ 0 & 0 & 1 \end{pmatrix}$$

The Jacobian of the relation is

$$J = 3(X_1X_2)^2$$

which is always positive for all coordinates X_1 and X_2. The above relation describes a motion. The inverse of the deformation gradient

$$(\mathbf{F}^{-1}) = \begin{pmatrix} 2/(3X_1X_2) & -1/(3X_2^2) & 0 \\ -1/(3X_1^2) & 2/(3X_1X_2) & 0 \\ 0 & 0 & 1 \end{pmatrix}$$

The corresponding Lagrangian and Eulerian strain tensors are as follows.
The Lagrangian displacement tensor $\mathbf{H}$, $\mathbf{H} = \mathbf{F} - \mathbf{I}$

$$(\mathbf{H}) = (\mathbf{F} - \mathbf{I}) = \begin{pmatrix} 2X_1X_2 - 1 & (X_1)^2 & 0 \\ (X_2)^2 & 2X_1X_2 - 1 & 0 \\ 0 & 0 & 0 \end{pmatrix}$$

and the Eulerian displacement tensor $\hat{\mathbf{H}}$, $\hat{\mathbf{H}} = \mathbf{I} - \mathbf{F}^{-1}$

$$(\hat{\mathbf{H}}) = (\mathbf{I} - \mathbf{F}^{-1}) = \begin{pmatrix} 1 - 2/(3X_1X_2) & 1/(3X_2^2) & 0 \\ 1/(3X_1^2) & 1 - 2/(3X_1X_2) & 0 \\ 0 & 0 & 0 \end{pmatrix}$$

6.3   Section 2.3

Example 2.3.1

We refer back to the St. Venant Torsion discussed in the previous example, where
the relation between both material and spatial coordinates is given as follows:

$$x_1 = X_1 - \frac{\theta}{l}X_2X_3$$
$$x_2 = X_2 + \frac{\theta}{l}X_3X_1$$
$$x_3 = X_3 + \psi(X_1, X_2)$$

The displacement vector u_i can be obtained directly using the equation

$$u_i = x_i - X_i$$

which produces

$$u_1 = -\frac{\theta}{l}X_2X_3$$
$$u_2 = \frac{\theta}{l}X_3X_1$$
$$u_3 = \psi(X_1, X_2)$$

the components of the displacement vector. From here, we can easily obtain the
linearized Lagrangian strain tensor, whose components are

$$\partial u_1/\partial X_1 = 0$$

$$\partial u_1/\partial X_2 = -\frac{\theta}{l}X_3$$

$$\partial u_1/\partial X_3 = -\frac{\theta}{l}X_2$$

$$\partial u_2/\partial X_1 = -\frac{\theta}{l}X_3$$

$$\partial u_2/\partial X_2 = 0$$

$$\partial u_2/\partial X_3 = \frac{\theta}{l}X_1$$

$$\partial u_3/\partial X_1 = \partial\psi(X_1, X_2)/\partial X_1$$

$$\partial u_3/\partial X_2 = \partial\psi(X_1, X_2)/\partial X_2$$

$$\partial u_3/\partial X_3 = 0$$

Rearranging the components to get the displacement gradient and its transpose, we have then

$$(\mathbf{H}) = \partial u_i/\partial X_j = \begin{pmatrix} 0 & -(\theta/l)X_3 & -(\theta/l)X_2 \\ (\theta/l)X_3 & 0 & (\theta/l)X_1 \\ \partial\psi(X_1, X_2)/\partial X_1 & \partial\psi(X_1, X_2)/\partial X_2 & 0 \end{pmatrix}$$

and

$$\mathbf{H} + \mathbf{H}^{\mathrm{T}} = \begin{pmatrix} 0 & 0 & \partial\psi/\partial X_1 - (\theta/l)X_2 \\ 0 & 0 & \partial\psi/\partial X_2 + (\theta/l)X_1 \\ \partial\psi/\partial X_1 - (\theta/l)X_2 & \partial\psi/\partial X_2 + (\theta/l)X_1 & 0 \end{pmatrix}$$

Therefore, the linearized Lagrangian strain has the following form:

$$(\hat{\mathbf{E}}) = \frac{1}{2}\begin{pmatrix} 0 & 0 & \partial\psi/\partial X_1 - (\theta/l)X_2 \\ 0 & 0 & \partial\psi/\partial X_2 + (\theta/l)X_1 \\ \partial\psi/\partial X_1 - (\theta/l)X_2 & \partial\psi/\partial X_2 + (\theta/l)X_1 & 0 \end{pmatrix}$$

We note that the Lagrangian strain tensor does not depend on the X_3 coordinate, i.e. the beam's axis. It solely depends on the cross-sectional coordinates, X_1 and X_2. It tells us that the deformation of the beam during the St. Venant torsion occurs only to the cross section of the beam and it is the same along the beam axis. The deformation that we have here is the rotation of the cross section as well the warping of the cross-sectional area.

Problem 2.3.1

Verify that the transformation below represents a motion

$$x_1 = X_1 e^t + X_3(e^t - 1)$$

$$x_2 = X_2 e^{-t} + X_3(e^t - e^{-t})$$

$$x_3 = X_3$$

If it is a motion, find the deformation gradient **F**. Determine the corresponding Lagrangian and Eulerian strain tensors.

Solution

The Lagrangian strain tensor **E** is calculated from the following.

Problem 2.3.2

As a result of displacement, the particles (X_1, X_2, X_3) of a body are situated at the points with the coordinates

$$x_1 = X_1 + \epsilon X_1, x_2 = X_2, x_3 = X_3$$

($\epsilon = $ const) with reference to a spatial Cartesian coordinate system x_i. What happens to the material lines with both ends initially at $(0,0,0)$ and $(1,0,0)$ and $0,0,0)$ and $(0,1,0)$ respectively following the motion described by the equations above? Consider the two cases: $\epsilon > 0$ and $|1 < \epsilon < 0$. Explain the results. Determine the corresponding Lagrangian and Eulerian strain tensors of this motion.

Solution

The material line $(0, 0, 0) - (1, 0, 0)$ will experience an elongation along the x_1 axis. After the motion, it becomes $(0, 0, 0) - (1 + \epsilon, 0, 0)$. The material line $(0, 0, 0) - (0, 1, 0)$ does not experience any deformation. At the end of the motion, it maintains the initial configuration, i.e. $(0, 0, 0) - (0, 1, 0)$.

Problem 2.3.3

Given is the following motion:

$$x_1 = X_1 \cos \frac{\pi}{4} t + X_2 \sin \frac{\pi}{4} t$$

$$x_2 = -X_1 \sin \frac{\pi}{4} t + X_2 \cos \frac{\pi}{4} t$$

$$x_3 = X_3$$

1. Find the position of a material line represented by a pair of Cartesian coordinates A(0, 1, 0), B(0, 2, 0) initially, after the time $t = 1$ lapses.
2. Calculate the length of this material line before the motion and after $t = 1$ motion.
3. Calculate the deformation gradient $\mathbf{F}$, and the right and left Green tensors, $\mathbf{C}$, $\mathbf{B}$ and
4. Show that the Lagrangian and Eulerian strain tensors, $\mathbf{E}$, $\mathbf{e}$, for this motion are null tensors.

Explain the results.

Solution

1. At the beginning point A is at the position (0,1,0), therefore the Lagrange coordinates for point A are

$$X_1 = 0, X_2 = 1, X_3 = 0$$

After $t = 1$, the new position of point A is given by the motion

$$x_1 = 0 \times \cos \frac{\pi}{4} + 1 \times \sin \frac{\pi}{4} = \frac{\sqrt{2}}{2}$$

$$x_2 = -0 \times \sin \frac{\pi}{4} + 1 \times \cos \frac{\pi}{4} = \frac{\sqrt{2}}{2}$$

$$x_3 = 0$$

Point B is initially at the position (0,2,0); the Lagrange coordinates for point B are

$$X_1 = 0, X_2 = 2, X_3 = 0$$

After $t = 1$, point B takes a new position

$$x_1 = 0 \times \cos \frac{\pi}{4} + 2 \times \sin \frac{\pi}{4} = \sqrt{2}$$

$$x_2 = -0 \times \sin\frac{\pi}{4} + 2 \times \cos\frac{\pi}{4} = \sqrt{2}$$

$$x_3 = 0$$

2. The length of the material line AB is 1 initially and after the motion it is

$$\sqrt{0^2 + (2 - 1)^2 + 0^2} = 1$$

There is no change in the length of the material line.
3. The deformation gradient **F** is calculated from the above-defined motion

$$(\mathbf{F}) = \partial x_i/\partial X_j = \begin{pmatrix} \cos\frac{\pi}{4}t & \sin\frac{\pi}{4}t & 0 \\ -\sin\frac{\pi}{4}t & \cos\frac{\pi}{4}t & 0 \\ 0 & 0 & 1 \end{pmatrix}$$

and the corresponding right and left Cauchy–Green tensors, **C**, **B** are

$$(\mathbf{C}) = (\mathbf{F} \cdot \mathbf{F}^T) = \begin{pmatrix} 1 & 0 & 0 \\ 0 & 1 & 0 \\ 0 & 0 & 1 \end{pmatrix} = \mathbf{I} = (\mathbf{F}^T \cdot \mathbf{F}) = (\mathbf{B})$$

The deformation gradient of the motion described above is an orthogonal tensor with det **F** = 1. Therefore
4. Lagrangian **E** and **e** and Eulerian strain tensor by definition are null tensors. *q.e.d*

The motion equations given above describe the pure rotational motion and no deformation involves.

Problem 2.3.4

For both homogeneous deformations described below

$$x_1 = X_1 + \kappa X_2; \quad x_2 = X_2; \quad x_3 = X_3$$

and

$$x_1 = \psi X_1; \quad x_2 = \frac{1}{\psi}X_2; \quad x_3 = X_3$$

obtain the corresponding deformation gradients. Show that they have the same strain tensors if $\kappa = |\psi - \frac{1}{\psi}|$.

Solution

The deformation gradient of the first homogeneous deformation is obtained as

$$(\mathbf{F_1}) = \begin{pmatrix} 1 & \kappa & 0 \\ 0 & 1 & 0 \\ 0 & 0 & 1 \end{pmatrix}$$

and the deformation gradient of the second deformation is calculated as

$$(\mathbf{F_2}) = \begin{pmatrix} \psi & 0 & 0 \\ 0 & 1/\psi & 0 \\ 0 & 0 & 1 \end{pmatrix}$$

The right Cauchy–Green strain tensors for both deformations are

$$(\mathbf{C_1}) = (\mathbf{F_1} \cdot \mathbf{F_1}^T) = \begin{pmatrix} 1 & \kappa & 0 \\ \kappa & \kappa^2+1 & 0 \\ 0 & 0 & 1 \end{pmatrix}$$

$$(\mathbf{F_2}) = \begin{pmatrix} \psi & 0 & 0 \\ 0 & 1/\psi & 0 \\ 0 & 0 & 1 \end{pmatrix}$$

The right Cauchy–Green strain tensors for both deformations are

$$(\mathbf{C_2}) = (\mathbf{F_2} \cdot \mathbf{F_2}^T) = \begin{pmatrix} \psi^2 & 0 & 0 \\ 0 & (1/\psi)^2+1 & 0 \\ 0 & 0 & 1 \end{pmatrix}$$

By equating both right Cauchy–Green tensors $bf C_1$ and $bf C_2$, we have

$$\psi^2 = 1$$

$$\kappa = 0$$

$$\kappa^2 + 1 = (1/\psi)2$$

By substituting $\kappa = |\psi - 1/\psi|$ in the third equation, we obtain the second and the first equations. Hence, we verify the condition $\kappa = |\psi - 1/\psi|$.

6.4   Section 2.4

Example 2.4.1

A body is rotated in three-dimensional space by an orthogonal tensor $\mathbf{R}$

$$\mathbf{R} = (1/3) \begin{pmatrix} 1 & 2 & 2 \\ 2 & 0 & -2 \\ -2 & 2 & -1 \end{pmatrix}$$

Verify that the tensor $\mathbf{R}$ is a proper orthogonal tensor and find the axis of rotation and corresponding rotation angle.

Solution

The determinant of $\mathbf{R}$ can be easily calculated and obtained.

$$\det \mathbf{R} = +1$$

Another property of an orthogonal tensor is

$$\mathbf{R} \cdot \mathbf{R}^{\mathrm{T}} = \mathbf{R}^{\mathrm{T}} \cdot \mathbf{R} = \mathbf{I}$$

We use Falk's matrix scheme to verify this property

$$(1/9) \begin{pmatrix} & & & 1 & 2 & 2 \\ & & & 2 & 0 & -2 \\ & & & -2 & 2 & -1 \\ 1 & 2 & -2 & 1 & 0 & 0 \\ 2 & 0 & 2 & 0 & 1 & 0 \\ 2 & -2 & -1 & 0 & 0 & 1 \end{pmatrix}$$

or

$$(1/9) \begin{pmatrix} & & & 1 & 2 & -2 \\ & & & 2 & 0 & 2 \\ & & & 2 & -2 & -1 \\ 1 & 2 & 2 & 1 & 0 & 0 \\ 2 & 0 & -2 & 0 & 1 & 0 \\ -2 & 2 & -1 & 0 & 0 & 1 \end{pmatrix}$$

The rotation axis can be obtained by determining the eigenvector of $\mathbf{R}$ corresponding to its real eigenvalues. After some calculations, the eigenvalues of $\mathbf{R}$ are

$$1, 1/3(-1 + 2i\sqrt{2}), 1/3(-1 - 2i\sqrt{2})$$

Since 1 is the only real eigenvalue, we determine the eigenvector associated with the eigenvalue equals 1. Again using symbolic software, we obtain the corresponding eigenvector to the real eigenvalue of $\mathbf{R}$ as follows:

$$\mathbf{v} = (1, 1, 0)$$

To find the rotation angle, we select one arbitrary vector and rotate this vector using the rotation tensor $\mathbf{R}$. The scalar product between the selected vector and the resulting vector will give us the required rotation angle. We select the start vector $\mathbf{a}$ as $\mathbf{a} = (1/\sqrt{3}(1, 1, 1)$. Then this vector is rotated by $\mathbf{R}$ to produce $\mathbf{b}$ according to

$$\mathbf{b} = \mathbf{R} \cdot \mathbf{a} = (1/3)(1/\sqrt{3}) \begin{pmatrix} 1 & 2 & 2 \\ 2 & 0 & -2 \\ -2 & 2 & -1 \end{pmatrix} (1, 1, 1)$$

$$= \frac{1}{9\sqrt{3}}(5, 1, -1)$$

The rotation angle ϕ can be obtained from the scalar product of these two vectors via the following equation for the scalar product, which produces

$$\mathbf{a} \cdot \mathbf{b} = \sqrt{(\mathbf{a} \cdot \mathbf{a})}\sqrt{(\mathbf{b} \cdot \mathbf{b})} \cos \phi$$

or

$$\cos \phi = \frac{\mathbf{a} \cdot \mathbf{b}}{\sqrt{(\mathbf{a} \cdot \mathbf{a})}\sqrt{(\mathbf{b} \cdot \mathbf{b})}}$$

With $\mathbf{a} \cdot \mathbf{b} = 5/3$, $\mathbf{a} \cdot \mathbf{a} = 3$ and $\mathbf{b} \cdot \mathbf{b} = 3$, the angle ϕ equals

$$\phi = \arccos(5/9)$$

Problem 2.4.1

Given is the plane motion described below

$$x_1 = 2X_1 + X_3; \quad x_2 = X_2; \quad x3 = X_1 + 3X_3$$

Find the associate right and left stretch tensors of this motion.

Solution

The deformation gradient $\mathbf{F}$ of this motion is

$$(\mathbf{F}) = \begin{pmatrix} 2 & 0 & 1 \\ 0 & 1 & 0 \\ 1 & 0 & 3 \end{pmatrix}$$

The right Cauchy–Green tensor $\mathbf{C} = \mathbf{F} \cdot \mathbf{F}^{\mathsf{T}}$ is

$$(\mathbf{C}) = \begin{pmatrix} 2 & 0 & 1 \\ 0 & 1 & 0 \\ 1 & 0 & 3 \end{pmatrix} = (\mathbf{U}^2)$$

The calculated eigenvalues or the principal values of $\mathbf{C} = (\mathbf{U}^2)$ are

$$U_I^2 = 5/2(3 + \sqrt{5}),\ U_{II}^2 = 5/2(3 - \sqrt{5}),\ U_{III}^2 = 1$$

and the corresponding eigenvectors or the principal directions

$$\mathbf{n_I} = (1/2(-1 + \sqrt{5}), 0, 1),\ \mathbf{n_{II}} = (1/2(-1 - \sqrt{5}), 0, 1),\ \mathbf{n_{III}} = (0, 1, 0)$$

With the principal values, the spectral form of the right Cauchy–Green tensor is expressed as follows:

$$(\mathbf{C}) = \begin{pmatrix} 5/2(3 + \sqrt{5}) & 0 & 0 \\ 0 & 5/2(3 - \sqrt{5}) & 1 \\ 0 & 0 & 1 \end{pmatrix} = (\mathbf{U}^2)$$

Therefore, the spectral form of the right stretch tensor is

$$(\mathbf{U}) = \begin{pmatrix} \sqrt{5/2(3 + \sqrt{5})} & 0 & 0 \\ 0 & \sqrt{5/2(3 - \sqrt{5})} & 0 \\ 0 & 0 & 1 \end{pmatrix} = \begin{pmatrix} 3.62 & 0 & 0 \\ 0 & 1.38 & 0 \\ 0 & 0 & 1 \end{pmatrix}$$

The right stretch tensor can be transformed back to the initial basis system; for that purpose, we construct the transformation matrix from the principal directions

$$\mathbf{T} = (\mathbf{n_I}, \mathbf{n_{II}}, \mathbf{n_{III}})^{T}$$

$$(\mathbf{T}) = \begin{pmatrix} 1/2(-1 + \sqrt{5}) & 1/2(-1 - \sqrt{5}) & 0 \\ 0 & 0 & 1 \\ 1 & 1 & 0 \end{pmatrix} = \begin{pmatrix} 0.62 & -1.62 & 0 \\ 0 & 0 & 1 \\ 1 & 1 & 0 \end{pmatrix}$$

Then the right stretch tensor $\mathbf{U}$ in the initial basis system is obtained from the following transformation:

$$(\mathbf{U}) = \begin{pmatrix} 0.62 & -1.62 & 0 \\ 0 & 0 & 1 \\ 1 & 1 & 0 \end{pmatrix} \begin{pmatrix} 3.62 & 0 & 0 \\ 0 & 1.38 & 0 \\ 0 & 0 & 1 \end{pmatrix} \begin{pmatrix} 0.62 & 0 & 1 \\ -1.62 & 0 & 1 \\ 0 & 1 & 0 \end{pmatrix}$$

After applying Falk's scheme (refer to Appendix V) for matrix multiplication, we get

$$(\mathbf{U}) = \begin{pmatrix} 5.87 & 0 & 0 \\ 0 & 1 & 1 \\ 1.38 & 1 & 5 \end{pmatrix}$$

the right stretch tensor expressed based on the initial basis system. The calculation for the left stretch tensor $\mathbf{V}$ can be done in a similar manner.

Problem 2.4.2

Apply the polar decomposition theorem to the shear motion described below

$$x_1 = X_1 + \kappa X_2; \quad x_2 = \kappa X_1 + X_2; \quad x_3 = X_3$$

Solution

Problem 2.4.3

Find the relationship between the right and left Cauchy–Green tensors, $\mathbf{C}$ and $\mathbf{B}$.

Solution

From

$$\mathbf{C} = \mathbf{F}^{\mathrm{T}} \cdot \mathbf{F}, \quad \mathbf{B} = \mathbf{F} \cdot \mathbf{F}^{\mathrm{T}}$$

we have

$$\mathbf{C} = \mathbf{U}^2 = \mathbf{F}^{\mathrm{T}} \cdot \mathbf{F}, \quad \mathbf{B} = \mathbf{V}^2 = \mathbf{F}^{\mathrm{T}} \cdot \mathbf{F}$$

Now

$$\mathbf{U}^2 = \mathbf{F}^{\mathrm{T}} \cdot \mathbf{F} = (\mathbf{V} \cdot \mathbf{R})^{\mathrm{T}} \cdot (\mathbf{V} \cdot \mathbf{R})$$

or

$$\mathbf{U}^2 = \mathbf{F}^{\mathrm{T}} \cdot \mathbf{F} = (\mathbf{R}^{\mathrm{T}} \cdot \mathbf{V}^{\mathrm{T}}) \cdot (\mathbf{V} \cdot \mathbf{R})$$

Since the tensor $\mathbf{U}$ is symmetrical, we have then

$$\mathbf{U}^2 = \mathbf{F}^T \cdot \mathbf{F} = (\mathbf{R}^T \cdot \mathbf{V}^2) \cdot \mathbf{R})$$

or we have the sought relation

$$\mathbf{C} = \mathbf{R}^T \cdot \mathbf{B} \cdot \mathbf{R}$$

6.5 Section 2.5

Example 2.5.1

In an experiment for the determination of plane strain, a 45° strain gauge rosette is used. From the measurement, it is found that

$$\epsilon_{0°}^* = 100\mu, \ \epsilon_{45°}^* = 50\mu, \ \epsilon_{90°}^* = -50\mu$$

Determine all the components of the plane strain tensor.

Solution

Use this equation

$$\epsilon_{11}^* = \epsilon_{11}(\cos\theta)^2 + \epsilon_{22}(\sin\theta)^2 + 2\epsilon_{12}\sin\theta\cos\theta$$

for ϵ_{11}^* at three different angles of 0°, 45° and 90° and get

$$\epsilon_{0°}^* = \epsilon_{11} + 0 + 0 = 100$$

$$\epsilon_{90°}^* = 0 + \epsilon_{22} + 0 = -50$$

$$\epsilon_{45°}^* = \epsilon_{11}(1/2) + \epsilon_{22}(1/2) + \epsilon_{12} = 100$$

We have then

$$\epsilon_{11} = 100\mu, \ \epsilon_{22} = -50\mu, \ \epsilon_{12} = 75\mu$$

The principal values of this strain can be obtained using Mohr's circle.

Example 2.5.2

If **T** is any tensor of second-order, show that it obeys its own characteristic equation

$$\mathbf{T}^3 - I_T\mathbf{T}^2 + II_T\mathbf{T} - III_T\mathbf{I} = \mathbf{0}$$

I_T, II_T and III_T are the first, second and third invariant of **T**.

Solution

Let λ_1, λ_2 and λ_3 be the eigenvalues of the tensor **T**. They satisfy the characteristic equation automatically, i.e.

$$\lambda_1^3 - I_T\lambda_1^2 + II_T\lambda_1 - III_T = 0$$

$$\lambda_2^3 - I_T\lambda_2^2 + II_T\lambda_2 - III_T = 0$$

$$\lambda_3^3 - I_T\lambda_3^2 + II_T\lambda_3 - III_T = 0$$

Rearranging together in a form of a matrix

$$\begin{pmatrix} \lambda_1^3 & 0 & 0 \\ 0 & \lambda_2^3 & 0 \\ 0 & 0 & \lambda_3^3 \end{pmatrix} - I_T \begin{pmatrix} \lambda_1^2 & 0 & 0 \\ 0 & \lambda_2^2 & 0 \\ 0 & 0 & \lambda_3^2 \end{pmatrix} - II_T \begin{pmatrix} \lambda_1 & 0 & 0 \\ 0 & \lambda_2 & 0 \\ 0 & 0 & \lambda_3 \end{pmatrix}$$

$$- III_T \begin{pmatrix} 1 & 0 & 0 \\ 0 & 1 & 0 \\ 0 & 0 & 1 \end{pmatrix} = \begin{pmatrix} 0 & 0 & 0 \\ 0 & 0 & 0 \\ 0 & 0 & 0 \end{pmatrix}$$

One notes that the matrix

$$\begin{pmatrix} \lambda_1 & 0 & 0 \\ 0 & \lambda_2 & 0 \\ 0 & 0 & \lambda_3 \end{pmatrix}$$

is a spectral matrix of the tensor **T**, denoted as $\hat{\mathbf{T}}$, so that we can rewrite the above matrix-form of the characteristic equation as

$$\hat{\mathbf{T}}^3 - I_T\hat{\mathbf{T}}^2 + II_T\hat{\mathbf{T}} - III_T\mathbf{I} = \mathbf{0}$$

A tensor of any order can be transformed to the same tensor based on different basis systems via an orthogonal transformation. In this case, a spectral matrix of tensor **T** can be transformed back to any other form using an appropriate orthogonal matrix, **Q** with the property $\mathbf{Q} \cdot \mathbf{Q}^T = \mathbf{Q}^T \cdot \mathbf{Q} = \mathbf{I}$ and $\det \mathbf{Q} = +1$. Now using the orthogonal transformation, we transform the spectral form of the tensor $\mathbf{T}$ to any other form through

$$\mathbf{Q} \cdot \hat{\mathbf{T}} \cdot \mathbf{Q}^T = \mathbf{T}$$

$$\mathbf{Q} \cdot \hat{\mathbf{T}}^2 \cdot \mathbf{Q}^T = (\mathbf{Q} \cdot \hat{\mathbf{T}} \cdot \mathbf{Q}^T) \cdot (\mathbf{Q} \cdot \hat{\mathbf{T}} \cdot \mathbf{Q}^T) = \mathbf{T}^2$$

$$\mathbf{Q} \cdot \hat{\mathbf{T}}^3 \cdot \mathbf{Q}^T \cdot \mathbf{Q} \cdot \hat{\mathbf{T}}^3 \cdot \mathbf{Q}^T \cdot \mathbf{Q} \cdot \hat{\mathbf{T}}^3 \cdot \mathbf{Q}^T = \mathbf{T}^3$$

Therefore, we have

$$\mathbf{T}^3 - I_T \mathbf{T}^2 + II_T \mathbf{T} - III_T \mathbf{I} = \mathbf{0}$$

Q.e.d

Example 2.5.3—Isotropic Tensor Function

Based on the Cayley–Hamilton theorem, any tensor-valued tensor function $\mathbf{F}(\mathbf{A})$ can be expressed as

$$\mathbf{F}(\mathbf{A}) = \phi_0 \mathbf{I} + \phi_1 \mathbf{A} + \phi_2 \mathbf{A}^2$$

Whereby ϕ_0, ϕ_1, ϕ_2 are scalar-valued function of invariants of the tensor $\mathbf{A}$. These functions are also called the integrity basis of the isotropic tensor functions.

If $\mathbf{U}$ is the right stretch tensor, then another strain measure describing the deformation, the natural logarithm of the right stretch tensor, sometimes called the Hencky strain tensor, can be expressed as an isotropic tensor function

$$\ln(\mathbf{U}) = \phi_0 \mathbf{I} + \phi_1 \mathbf{U} + \phi_2 \mathbf{U}^2$$

Determine the scalar value functions ϕ_0, ϕ_1, ϕ_2 for a complete description of the Hencky strain tensor.

Solution

The determination of the integrity basis ϕ_0, ϕ_1, ϕ_2 of the isotropic tensor function $\ln(\mathbf{U})$ can be done in the principle space, i.e. based on the principal directions or eigenvectors. Assuming that U_I, $U_{I}I$ and $U_{I}II$ are the principal values of the right stretch tensor $\mathbf{U}$, then the associated Hencky strain tensor can be written as

$$\begin{pmatrix} \ln U_I & 0 & 0 \\ 0 & \ln U_{II} & 0 \\ 0 & 0 & \ln U_{III} \end{pmatrix} = \phi_0 \begin{pmatrix} 1 & 0 & 0 \\ 0 & 1 & 0 \\ 0 & 0 & 1 \end{pmatrix} + \phi_1 \begin{pmatrix} U_I & 0 & 0 \\ 0 & U_{II} & 0 \\ 0 & 0 & U_{III} \end{pmatrix}$$

$$+ \phi_2 \begin{pmatrix} U_I^2 & 0 & 0 \\ 0 & U_{II}^2 & 0 \\ 0 & 0 & U_{III}^2 \end{pmatrix}$$

Now there are three equations with the three unknown integrity bases:

$$\begin{pmatrix} 1 & U_I & U_I^2 \\ 1 & U_{II} & U_{II}^2 \\ 1 & U_{III} & U_{III}^2 \end{pmatrix} \begin{pmatrix} \phi_0 \\ \phi_1 \\ \phi_2 \end{pmatrix} = \begin{pmatrix} \ln U_I \\ \ln U_{II} \\ \ln U_{III} \end{pmatrix}$$

or

$$\begin{pmatrix} \phi_0 \\ \phi_1 \\ \phi_2 \end{pmatrix} = \begin{pmatrix} 1 & U_I & U_I^2 \\ 1 & U_{II} & U_{II}^2 \\ 1 & U_{III} & U_{III}^2 \end{pmatrix}^{-1} \begin{pmatrix} \ln U_I \\ \ln U_{II} \\ \ln U_{III} \end{pmatrix}$$

Problem 2.5.1

In the following is a spectral form of a strain tensor $\hat{\mathbf{T}}$ given as

$$\hat{\mathbf{T}} = \begin{pmatrix} 2 & 0 & 0 \\ 0 & 3 & 0 \\ 0 & 0 & -3 \end{pmatrix}$$

Perform an orthogonal transformation with the following orthogonal tensor $\mathbf{Q}$

$$\mathbf{Q} = \frac{1}{3} \begin{pmatrix} 1 & 2 & 2 \\ 2 & 1 & -2 \\ -2 & 2 & -1 \end{pmatrix}$$

Show that these are two different forms of the same strain tensor, and hence show that the expression

$$\mathbf{T}^3 - I_T \mathbf{T}^2 + II_T \mathbf{T} - III_T \mathbf{I} \tag{6.1}$$

is a zero tensor.

Solution

We start first with the orthogonal transformation of the spectral form of the strain tensor $\hat{\mathbf{T}}$ and obtain

$$\mathbf{T} = \mathbf{Q} \cdot \hat{\mathbf{T}} \cdot \mathbf{Q}^T$$

$$\mathbf{T} = \frac{1}{3}\begin{pmatrix} 1 & 2 & 2 \\ 2 & 1 & -2 \\ -2 & 2 & -1 \end{pmatrix} \cdot \begin{pmatrix} 2 & 0 & 0 \\ 0 & 3 & 0 \\ 0 & 0 & -3 \end{pmatrix} \cdot \frac{1}{3}\begin{pmatrix} 1 & 2 & 2 \\ 2 & 1 & -8 \\ 14 & -8 & 17 \end{pmatrix}^T$$

$$\mathbf{T} = \frac{1}{9}\begin{pmatrix} 2 & 22 & 14 \\ 22 & -1 & -2 \\ -2 & 2 & -1 \end{pmatrix}$$

These two matrices represent the same strain tensor as both $(\hat{\mathbf{T}})$ and $(\mathbf{T})$ have the same invariants $I_T = 2$, $II_T = -9$, $III_T = -18$. Substitute the obtained matrix of the tensor $\mathbf{T}$,

$$\frac{1}{9^3}\begin{pmatrix} 2 & 22 & 14 \\ 22 & -1 & -2 \\ -2 & 2 & -1 \end{pmatrix}^3 - (2)\frac{1}{9^2}\begin{pmatrix} 2 & 22 & 14 \\ 22 & -1 & -2 \\ -2 & 2 & -1 \end{pmatrix}^2 + (-9)\frac{1}{9}\begin{pmatrix} 2 & 22 & 14 \\ 22 & -1 & -2 \\ -2 & 2 & -1 \end{pmatrix}$$

$$-(-18)\begin{pmatrix} 1 & 0 & 0 \\ 0 & 1 & 0 \\ 0 & 0 & 1 \end{pmatrix}$$

$$\begin{pmatrix} 8/9 & 178/9 & 146/9 \\ 178/9 & -49/9 & -32/9 \\ 146/9 & -32/9 & 113/9 \end{pmatrix} - (2)\begin{pmatrix} 76/9 & -10/9 & 10/9 \\ -10/9 & 61/9 & 20/9 \\ 10/9 & 20/9 & 61/9 \end{pmatrix}$$

$$+(-9)\begin{pmatrix} 2/9 & 22/9 & 14/9 \\ 22/9 & -1/9 & -8/9 \\ 14/9 & -8/9 & 17/9 \end{pmatrix} - (-18)\begin{pmatrix} 1 & 0 & 0 \\ 0 & 1 & 0 \\ 0 & 0 & 1 \end{pmatrix}$$

$$= \begin{pmatrix} 0 & 0 & 0 \\ 0 & 0 & 0 \\ 0 & 0 & 0 \end{pmatrix}$$

Therefore, any tensor $\mathbf{T}$ satisfies its own characteristic equation automatically.

Problem 2.5.2

Calculate the value of the following isotropic tensor function:

$$\sin \begin{pmatrix} 1 & 3 & 2 \\ 3 & 1 & 5 \\ 2 & 5 & 2 \end{pmatrix}$$

Solution

Firstly, we calculate the principal values of the argument tensor

$$\begin{pmatrix} 1 & 3 & 2 \\ 3 & 1 & 5 \\ 2 & 5 & 2 \end{pmatrix} = \mathbf{A}$$

They are

$$A_I = 8.23,\ A_{II} = -3.74,\ A_{III} = -0.49$$

Now the spectral form of the tensor **A** is

$$\mathbf{A} = \begin{pmatrix} 8.23 & 0 & 0 \\ 0 & -3.74 & 0 \\ 0 & 0 & -0.49 \end{pmatrix}$$

Now the spectral form of the sine function of the tensor **A** can be obtained directly by applying the sine function to its spectral elements,

$$\mathbf{B} = \sin \begin{pmatrix} 8.23 & 0 & 0 \\ 0 & -3.74 & 0 \\ 0 & 0 & -0.49 \end{pmatrix}$$

$$= \begin{pmatrix} \sin 8.23 & 0 & 0 \\ 0 & \sin(-3.74) & 0 \\ 0 & 0 & \sin(-0.49) \end{pmatrix}$$

$$= \begin{pmatrix} 0.93 & 0 & 0 \\ 0 & 0.56 & 0 \\ 0 & 0 & -0.47 \end{pmatrix}$$

The transformation back with its spectral matrix, i.e. the matrix with its eigenvectors as elements, yields the tensor **B** expressed with respect to its initial basis system. The normalized and orthogonal eigenvectors of tensor **A** are

$$\mathbf{n_I} = (0.44, 0.63, 0.65), \quad \mathbf{n_{II}} = (0.24, -0.77, 0.59)$$

$$\mathbf{n_{III}} = (-0.87, 0.10, 0.49)$$

The transformation matrix $\mathbf{P}$ with the above orthonormalized eigenvectors as elements is

$$\mathbf{P} = \begin{pmatrix} 0.44 & 0.24 & -0.87 \\ 0.63 & -0.77 & 0.10 \\ 0.65 & 0.59 & 0.49 \end{pmatrix}$$

Now the tensor $\mathbf{B}$ can be expressed back in the initial basis system $\hat{\mathbf{B}}$ using the following transformation:

$$\hat{\mathbf{B}} = \mathbf{P} \cdot \mathbf{B} \cdot \mathbf{P^T}$$

$$\hat{\mathbf{B}} = \mathbf{P} \cdot \begin{pmatrix} 0.93 & 0 & 0 \\ 0 & 0.56 & 0 \\ 0 & 0 & -0.47 \end{pmatrix} \cdot \mathbf{P^T}$$

$$\hat{\mathbf{B}} = \begin{pmatrix} -0.14 & 0.19 & 0.54 \\ 0.19 & 0.69 & 0.10 \\ 0.54 & 0.10 & 0.47 \end{pmatrix}$$

6.6   Section 2.6

Example 2.6.1

Given is the following St. Venant torsion:

$$x_1 = X_1 - \frac{\theta}{l} X_2 X_3$$

$$x_2 = X_2 + \frac{\theta}{l} X_3 X_1$$

$$x_3 = X_3 + \psi(X_1, X_2)$$

Investigate the above motion if it deals with dilatation or distortion.

Solution

The displacement vector u_i can be obtained directly using the equation

$$u_i = x_i - X_i$$

which produces

$$u_1 = -\frac{\theta}{l} X_2 X_3$$

$$u_2 = \frac{\theta}{l} X_3 X_1$$

$$u_3 = \psi(X_1, X_2)$$

the components of the displacement vector. The corresponding displacement gradient **H**

$$(\mathbf{H}) = \partial u_i / \partial X_j = \begin{pmatrix} 0 & -(\theta/l)X_3 & -(\theta/l)X_2 \\ (\theta/l)X_3 & 0 & (\theta/l)X_1 \\ \partial\psi(X_1, X_2)/\partial X_1 & \partial\psi(X_1, X_2)/\partial X_2 & 0 \end{pmatrix}$$

and

$$\mathbf{H} + \mathbf{H}^{\mathrm{T}} = \begin{pmatrix} 0 & 0 & \partial\psi/\partial X_1 - (\theta/l)X_2 \\ 0 & 0 & \partial\psi/\partial X_2 + (\theta/l)X_1 \\ \partial\psi/\partial X_1 - (\theta/l)X_2 & \partial\psi/\partial X_2 + (\theta/l)X_1 & 0 \end{pmatrix}$$

Therefore, the linearized Lagrangian strain has the following form:

$$(\mathbf{E}) = \frac{1}{2} \begin{pmatrix} 0 & 0 & \partial\psi/\partial X_1 - (\theta/l)X_2 \\ 0 & 0 & \partial\psi/\partial X_2 + (\theta/l)X_1 \\ \partial\psi/\partial X_1 - (\theta/l)X_2 & \partial\psi/\partial X_2 + (\theta/l)X_1 & 0 \end{pmatrix}$$

The hydrostatic part of the linearized Green (Lagrange) strain tensor is 0 because

$$tr(\mathbf{E}) = 0$$

So in the St. Venant torsion there is no dilatation, i.e. no change in volume. Thus, the deviatoric part $\hat{\mathbf{E}}'$ is **E** itself

$$\hat{\mathbf{E}}' = \frac{1}{2} \begin{pmatrix} 0 & 0 & \partial\psi/\partial X_1 - (\theta/l)X_2 \\ 0 & 0 & \partial\psi/\partial X_2 + (\theta/l)X_1 \\ \partial\psi/\partial X_1 - (\theta/l)X_2 & \partial\psi/\partial X_2 + (\theta/l)X_1 & 0 \end{pmatrix}$$

So the deformation of the body during the St. Venant torsion is merely a change in the shape of the cross section due to shear and warping.

Problem 2.6.1

On one particular point of a deformable body, the displacement gradient is measured to have the following form:

$$(\nabla \mathbf{u}) = \begin{pmatrix} 1 & -3 & 2 \\ 2 & 1 & 5 \\ 0 & -3 & 2 \end{pmatrix} \times 10^{-3}$$

1. Find the infinitesimal strain and spin tensors.
2. Calculate the spherical and deviatoric parts of the infinitesimal strain tensor.
3. Calculate the associated eigenvalues and eigenvectors of the infinitesimal strain tensor.

6.7 Section 2.7

Example 2.7.1

The motion of a deformable body is given by the following relations:

$$x_1 = X_1$$
$$x_2 = (1/2)[(X_2 + X_3) \exp(\alpha)t] + (1/2)[(X_2 - X_3) \exp(-\alpha)t]$$
$$x_3 = (1/2)[(X_2 + X_3) \exp(\alpha)t] - (1/2)[(X_2 - X_3) \exp(-\alpha)t]$$

Determine the velocity field of this object in Lagrangian or material form. Calculate its velocity gradient tensor **L**.

Solution

The deformation gradient **F** of this motion is

$$(\mathbf{F}) = \begin{pmatrix} 1 & 0 & 0 \\ 0 & (1/2)\exp(\alpha)t + (1/2)\exp(-\alpha)t & (1/2)\exp(\alpha)t - (1/2)\exp(-\alpha)t \\ 0 & (1/2)\exp(\alpha)t + (1/2)\exp(-\alpha)t & (1/2)\exp(\alpha)t + (1/2)\exp(-\alpha)t \end{pmatrix}$$

and the velocity field is

$$\dot{x}_1 = v_1 = 0$$
$$\dot{x}_2 = v_2 = (\alpha/2)[(X_2 + X_3) \exp(\alpha)t] - (\alpha/2)[(X_2 - X_3) \exp(-\alpha)t]$$
$$\dot{x}_3 = v_3 = (\alpha/2)[(X_2 + X_3) \exp(\alpha)t] + (\alpha/2)[(X_2 - X_3) \exp(-\alpha)t]$$

The deformation velocity tensor $\dot{\mathbf{F}}$ is obtained by differentiating $\mathbf{F}$ with respect to time,

$$(\dot{\mathbf{F}}) = \begin{pmatrix} 1 & 0 & 0 \\ 0 & (\alpha/2)\exp(\alpha)t - (\alpha/2)\exp(-\alpha)t & (\alpha/2)\exp(\alpha)t + (\alpha/2)\exp(-\alpha)t \\ 0 & (\alpha/2)\exp(\alpha)t - (\alpha/2)\exp(-\alpha)t & (\alpha/2)\exp(\alpha)t - (\alpha/2)\exp(-\alpha)t \end{pmatrix}$$

and as it was shown before that this is also the velocity gradient tensor $\mathbf{L}$.

Problem 2.7.1

The following relations describe the planar velocity field of a fluid motion:

$$v_1 = x_1$$
$$v_2 = 0$$
$$v_3 = \frac{x_3}{(1 + \kappa t)}$$

Determine the path and stream lines of the motion at the time t_0 and passing through the point P with the coordinates (P_1, P_2, P_3).

Solution

The path lines of the fluid motion are obtained by integrating the velocity field over time. In this case, we have for example

$$v_1 = dx_1/dt = x_1$$

$$dx_1/x_1 = dt$$

or

$$x_1 = A\exp(\alpha t)$$

$$x_2 = B$$

$$x_3 =$$

The stream lines can be obtained from the condition

$$\mathbf{v} \times d\mathbf{x} = \mathbf{0}$$

with $\mathbf{v} = (v_1, v_2, v_3)^T$ and $d\mathbf{x} = (dx_1, dx_2, dx_3)^T$

The integration of the equation derived from this condition with initial conditions $t = t_0$ delivers the general equation for stream lines. Adapting the general streamline equation to the point with coordinates (P_1, P_2, P_3) produces the special streamline equation.

6.8 Section 2.8

Example 2.8.1

Given is the following two-dimensional displacement field of a body.

$$u_1 = Ax_1x_2$$

$$u_2 = Bx_2x_2$$

where A and B are small constants. Determine

1. the components of linearized (small) strain,
2. whether the strains obey the compatibility condition.

Solution

For small strains, one starts from

$$\epsilon_{ij} = \frac{1}{2}\left(\frac{\partial u_i}{\partial x_j} + \frac{\partial u_j}{\partial x_i}\right) = \frac{1}{2}(u_{i,j} + u_{j,i})$$

and gets the linearized strain tensor

$$(\mathbf{E}) = \epsilon_{ij} = \frac{1}{2}\begin{pmatrix} 2Ax_2 & Ax_1 \\ Ax_1 & 4Bx_2 \end{pmatrix}$$

With the above tensor of small strain for the compatibility condition for the two-dimensional case, Eq. (2.23) is automatically fulfilled.

Problem 2.8.1

A square plate is a clamped support along the x- and y-axes and loaded with externally distributed load along its diagonal. From the measurement data taken by strain gauges, it is assumed that the components of the strain tensor take the following form:

$$\epsilon_{xx} = Ax^2y + y^3$$

and

$$\epsilon_{yy} = Bxy^2$$

Based on this assumption, determine the displacement field and the slip γ_{xy}. What is the condition for the fulfillment of the compatibility conditions?

Solution

From the given assumption for the strain

$$\epsilon_{xx} = \partial u_x/\partial x = Ax^2 y + y^3$$

and

$$\epsilon_{yy} = \partial u_y/\partial y = Bxy^2$$

we obtain by integration

$$u_x = 1/3 Ax^3 y + y^3 x + f(y)$$

and

$$u_y = 1/3 Bxy^3 + g(x)$$

Fitting to the boundary conditions, i.e. clamped along the x- and y-axes,

$$u_x(x, y = 0) = 0$$

$$u_y(x = 0, y) = 0$$

we have

$$f(y) = 0, g(x) = 0$$

The slip γ_{xy} is obtained from

$$\gamma_{xy} = \epsilon_{xy} + \epsilon_{yx} = \partial u_x/\partial y + \partial u_y/\partial x$$

$$= \frac{1}{3}(Ax^3 + 9y^2 + By^3)$$

We use Eq. (2.23) to check compatibility conditions

$$\frac{\partial^2 \epsilon_{xx}}{\partial^2 y} + \frac{\partial^2 \epsilon_{yy}}{\partial^2 x} = 2\frac{\partial^2 \epsilon_{xy}}{\partial x \partial y}$$

$$\frac{\partial^2 \epsilon_{xx}}{\partial^2 y} = 0$$

$$\frac{\partial^2 \epsilon_{yy}}{\partial^2 x} = 0$$

$$\frac{\partial^2 \epsilon_{xy}}{\partial x \partial y} = 0$$

The assumption that the component of the strain tensor $\epsilon_{xx} = \partial u_x/\partial x = Ax^2 y + y^3$ and $\epsilon_{yy} = \partial u_y/\partial y = Bxy^2$ complies with the compatibility condition.

6.9 Section 2.9

Example 2.9.1

Figure 2.8 shows a bounded plane is approximated by a quadrilateral with vertices $1: (-5, -1), 2: (-4, 2), 3: (-1, 1)$ and $4: (-2, -2)$. After deformation, the vertices displace to the position $1': (1, 0), 2': (3, 4), 3': (6.5, 1)$ and $4': (4.5, -2.5)$. Calculate the displacement field of the plane and compute the displacement vector of a point $P(-3.5, 1)$. Express the position of the point P in global coordinates x, y. Calculate the global deformation gradient $\mathbf{F}$ and identify the rotation and the stretching part in the deformation.

Solution

The positions of the vertices in the reference configuration are listed below

$$x_1 = -5, x_2 = -4, x_3 = -1, x_4 = -2$$

$$y_1 = -1, y_2 = 2, y_3 = 1, y_4 = -2$$

The displacement vector of the vertices or nodes are

$$u_1 = 6, u_2 = 7, u_3 = 7.5, u_4 = 6.5$$

$$v_1 = 1, v_2 = 2, v_3 = 0, v_4 = -0.5$$

In terms of global coordinates (x, y), the point $P(-3.5, 1)$ can be expressed as

$$x = 1/4(1-r)(1-s)x1 + 1/4(1+r)(1-s)x2 + 1/4(1+r)(1+s)x3 + 1/4(1-r)(1+s)x4$$

$$y = 1/4(1-r)(1-s)y1 + 1/4(1+r)(1-s)y2 + 1/4(1+r)(1+s)y3 + 1/4(1-r)(1+s)y4$$

Upon substitution, it produces

$$x_P = -0.17125, \; y_P = 1.1$$

The displacement field in terms of local coordinates (r, s) is given as follows:

$$u(r, s) = \frac{1}{4}(1-r)(1-s)u_1 + \frac{1}{4}(1+r)(1-s)u_2$$
$$+ \frac{1}{4}(1+r)(1+s)u_3 + \frac{1}{4}(1-r)(1+s)u_4$$

and

$$v(r, s) = \frac{1}{4}(1-r)(1-s)v_1 + \frac{1}{4}(1+r)(1-s)v_2$$
$$+ \frac{1}{4}(1+r)(1+s)v_3 + \frac{1}{4}(1-r)(1+s)v_4$$

On substitution of the individual displacement of the nodes, we have the displacement field of the element

$$u(r, s) = \frac{3}{2}(1-r)(1-s) + \frac{7}{4}(1+r)(1-s) + \frac{15}{8}(1+r)(1+s) + \frac{13}{8}(1-r)(1+s)$$

$$v(r, s) = \frac{1}{4}(1-r)(1-s) + \frac{1}{2}(1+r)(1-s) - \frac{1}{8}(1-r)(1+s)$$

For the displacement of the point $P(-3.5, 1)$ for which we set $r = -3.5$ and $s = 1$, we obtain

$$u(-3.5, 1) = \frac{3}{2}(1-(-3.5))(1-(1)) + \frac{7}{4}(1+(-3.5))(1-(1))$$
$$+ \frac{15}{8}(1+(-3.5))(1+(1))$$
$$+ \frac{13}{8}(1-(-3.5))(1+(1))$$
$$= 5.875$$

$$v(r, s) = \frac{1}{4}(1-(-3.5))(1-(1)) + \frac{1}{2}(1-(-3.5))(1-(1)) - \frac{1}{8}(1-(-3.5))(1-(1))$$
$$= -1.125$$

Therefore, the displacement of the material point P to P' is

$$\mathbf{u}_P = (5.875, -1.125)$$

From the displacement field obtained previously, the elements of the displacement gradient in local coordinate are calculated as follows:

$$\partial u/\partial r = -\frac{1}{4}(1-s)u_1 + \frac{1}{4}(1-s)u_2 + \frac{1}{4}(1+s)u_3 - \frac{1}{4}(1+s)u_4$$

$$\partial u/\partial s = -\frac{1}{4}(1-r)u_1 - \frac{1}{4}(1+r)u_2 + \frac{1}{4}(1+r)u_3 + \frac{1}{4}(1-r)u_4$$

$$\partial v/\partial r = -\frac{1}{4}(1-s)v_1 + \frac{1}{4}(1-s)v_2 + \frac{1}{4}(1+s)v_3 - \frac{1}{4}(1+s)v_4$$

$$\partial v/\partial s = -\frac{1}{4}(1-r)v_1 + \frac{1}{4}(1-r)v_2 + \frac{1}{4}(1+r)v_3 - \frac{1}{4}(1+r)v_4$$

One can obtain the global displacement gradient performing the coordinate transform between these two coordinate systems. We get first the global and local coordinate transformation:

$$\partial x/\partial r = -\frac{1}{4}(1-s)x_1 + \frac{1}{4}(1-s)x_2 + \frac{1}{4}(1+s)x_3 - \frac{1}{4}(1+s)x_4$$

$$\partial x/\partial s = -\frac{1}{4}(1-r)x_1 - \frac{1}{4}(1+r)x_2 + \frac{1}{4}(1+r)x_3 + \frac{1}{4}(1-r)x_4$$

$$\partial y/\partial r = -\frac{1}{4}(1-s)y_1 + \frac{1}{4}(1-s)y_2 + \frac{1}{4}(1+s)y_3 - \frac{1}{4}(1+s)y_4$$

$$\partial y/\partial s = -\frac{1}{4}(1-r)y_1 + \frac{1}{4}(1-r)y_2 + \frac{1}{4}(1+r)y_3 - \frac{1}{4}(1+r)y_4$$

Now after applying the chain rule, the global displacement gradient $\partial \mathbf{u}/\partial \mathbf{x}$ is given as

$$\partial \mathbf{u}/\partial \mathbf{x} = \partial \mathbf{u}/\partial \xi \cdot \partial \xi/\partial \mathbf{x} = \partial \mathbf{u}/\partial \xi \cdot (\partial \mathbf{x}/\partial \xi)^{-1}$$

or in matrix form

$$\begin{pmatrix} \partial u/\partial x & \partial u/\partial y \\ \partial v/\partial x & \partial v/\partial y \end{pmatrix} = \begin{pmatrix} \partial u/\partial r & \partial u/\partial s \\ \partial v/\partial r & \partial v/\partial s \end{pmatrix} \begin{pmatrix} \partial r/\partial x & \partial r/\partial y \\ \partial s/\partial x & \partial s/\partial y| \end{pmatrix}$$

$$= \begin{pmatrix} \partial u/\partial r & \partial u/\partial s \\ \partial v/\partial r & \partial v/\partial s \end{pmatrix} \begin{pmatrix} \partial x/\partial r & \partial x/\partial s \\ \partial y/\partial r & \partial y/\partial s \end{pmatrix}^{-1}$$

By substituting the numerical values used above, the matrix equation can be written as

$$\begin{pmatrix} \partial u/\partial x & \partial u/\partial y \\ \partial v/\partial x & \partial v/\partial y \end{pmatrix} = \begin{pmatrix} 0.5 & 0.25 \\ 0.25 & -3.94 \end{pmatrix} \begin{pmatrix} 6. & 4. \\ -1.33 & -1.33 \end{pmatrix}$$

$$\partial u/\partial x = \begin{pmatrix} 2.67 & 1.67 \\ 6.75 & 6.25 \end{pmatrix}$$

The global deformation gradient at the point $P(-3.5, 1)$ is then calculated from

$$\mathbf{F} = \mathbf{I} + \partial u/\partial x$$

$$\mathbf{F} = \begin{pmatrix} 3.67 & 1.67 \\ 6.75 & 7.25 \end{pmatrix}$$

Rotation or stretching of the element can be obtained from the polar decomposition of the deformation gradient,

$$\mathbf{F} = \mathbf{R} \cdot \mathbf{U} = \mathbf{V} \cdot \mathbf{R}$$

The left or right stretch tensor, $\mathbf{V}$ or $\mathbf{U}$, can be determined from

$$\mathbf{V}^2 = \mathbf{F} \cdot \mathbf{F}^T$$

or

$$\mathbf{U}^2 = \mathbf{F}^T \cdot \mathbf{F}$$

With the above-calculated deformation gradient, both stretch tensors are

$$\mathbf{U}^2 = \begin{pmatrix} 52.72 & 46.66 \\ 46.66 & 41.88 \end{pmatrix}$$

and

$$\mathbf{V}^2 = \begin{pmatrix} 9.89 & 28.43 \\ 28.43 & 84.71 \end{pmatrix}$$

As we can see, the right and left stretch tensors and their squares are not equal and the difference between these two stretch tensors is significantly small. This suggests that there is a rotation of the element in the deformation process besides the stretching and the rotation is relatively minimal. We leave it to the reader to find the associated rotation tensor $\mathbf{R}$ in this deformation.

Problem 2.9.1

A quadrilateral is originally with vertices at the points $1 : (2, 1), 2 : (-3, 3), 3 :$ $(-1, 4)$ and $4 : (-2, -2)$. After deformation, the vertices are displaced to the positions $1' : (4, 5), 2' : (6, 4), 3' : (5, -3)$ and $4' : (4, -2)$. Calculate the displacement field of the plane and express the position of the point $P(1, 1)$ of an undeformed quadrilateral in terms of the global coordinates x, y. Calculate the global deformation gradient $\mathbf{F}$ and the associated Lagrangian strain tensor $\mathbf{E}$.

Solution

The positions of the vertices in the reference configuration are listed below

$$x_1 = 2, x_2 = -3, x_3 = -1, x_4 = -2$$

$$y_1 = 1, y_2 = 3, y_3 = 4, y_4 = -2$$

The displacement vector of the vertices or nodes are

$$u_1 = 2, u_2 = 9, u_3 = 6, u_4 = 6$$

$$v_1 = 4, v_2 = 1, v_3 = -7, v_4 = 0$$

In terms of global coordinates (x, y), the point $P(1, 1)$ can be expressed as

$$x = 1/4(1 - r)(1 - s)x1 + 1/4(1 + r)(1 - s)x2 + 1/4(1 + r)(1 + s)x3 + 1/4(1 - r)(1 + s)x4$$

$$y = 1/4(1 - r)(1 - s)y1 + 1/4(1 + r)(1 - s)y2 + 1/4(1 + r)(1 + s)y3 + 1/4(1 - r)(1 + s)y4$$

Upon substitution, it produces

$$x_P = -1, y_P = 4$$

6.10   Section 3.2

Example 3.2.1

A cylindrical rod represented by a cylinder $x_1x_1 + x_2x_2 = 1$ is loaded with torsion moment so that the resulting shear stress takes the form $\sigma_{12} = 2\sigma$, where σ is a constant (refer to Fig. 3.2). The atmospheric pressure is assumed to be $\kappa\sigma$. Calculate traction vector **t** with respect to the tangential plane of the cylindrical rod at the point P(4/5, 3/5, 1). Calculate also the component vectors of **t** on that plane.

Solution

Due to the loading and atmospheric pressure, the state of stress in the rod represented by the Cauchy stress tensor **S** superposed by the hydrostatic (atmospheric) pressure and the torsion (Fig. 6.2),

$$(\mathbf{S}) = \sigma_{ij} = \kappa\sigma \begin{pmatrix} 1 & 0 & 0 \\ 0 & 1 & 0 \\ 0 & 0 & 1 \end{pmatrix} + \begin{pmatrix} 0 & 2\sigma & 0 \\ 2\sigma & 0 & 0 \\ 0 & 0 & 0 \end{pmatrix}$$

$$(\mathbf{S}) = \sigma_{ij} = \kappa\sigma \begin{pmatrix} 1 & 2/\kappa & 0 \\ 2/\kappa & 1 & 0 \\ 0 & 0 & 1 \end{pmatrix}$$

The gradient of the cylindrical surface of the rod at any point is obtained from

$$\mathbf{f} = \nabla(x_1 x_1 + x_2 x_2 = 1) = (2x_1, 2x_2, 0)$$

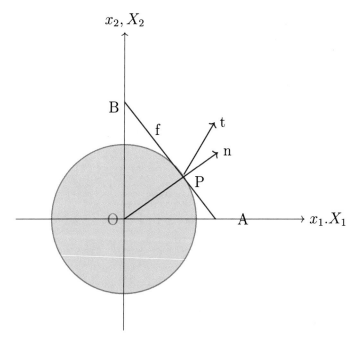

Fig. 6.2 Problem 3.1.2

For the point P(4/5,3/5,1), the gradient or slope vector is then

$$\mathbf{f} = (8/5, 6/5, 0)$$

A vector perpendicular to this gradient vector is then

$$\mathbf{g} = (6, -8, 0)$$

The normalized $\mathbf{g}$ is the normal vector to the above gradient vector; here

$$\mathbf{n} = 1/(100)(6, -8, 0)$$

Now we are to calculate the traction vector associated with the above normal vector, that is

$$\mathbf{t}/\kappa\sigma = \begin{pmatrix} & & & 6/100 \\ & & & -8/100 \\ & & & 0 \\ 1 & 2/\kappa & 0 & \frac{1}{100}(6 - 16/\kappa) \\ 2/\kappa & 1 & 0 & \frac{1}{100}(12 - 8/\kappa) \\ 0 & 0 & 1 & 0 \end{pmatrix}$$

which produces

$$\mathbf{t} = \frac{\kappa\sigma}{100}((6 - 16/\kappa), (12 - 8/\kappa), 0)$$

the sought traction vector. The component of the traction vector that is parallel to the normal vector is

$$\mathbf{t_n} = (\mathbf{t} \cdot \mathbf{n})\mathbf{n}$$

Therefore, the component of $\mathbf{t}$, that is on the plane, i.e. tangential components is

$$\mathbf{t_t} = \mathbf{t} - (\mathbf{t} \cdot \mathbf{n})\mathbf{n}$$

or by introducing a dyadic product $\otimes$,[1] we have

$$\mathbf{t_t} = (\mathbf{I} - (\mathbf{n} \otimes \mathbf{n})) \cdot \mathbf{t}$$

Problem 3.2.1

A homogeneous heavy piece of wood moves uniformly along a horizontal plane with a rough surface and the coefficient of dry friction $\mu_o = 0.35$. The area of the contact surface equals 0.012 m^2, and the mass of the wood block equals 10 kg. Find the stress vector on the plane under the wooden block.

---

[1] Refer to Appendix I

6.11 Section 3.3

Example 3.3.1

Given is the following Cauchy stress tensor:

$$(\mathbf{S}) = \sigma_{ij} = \begin{pmatrix} 1 & 2 & 1 \\ 2 & 3 & 0 \\ 1 & 0 & -2 \end{pmatrix} \text{ MPa}$$

Calculate all the invariants and hence find the principal values of the tensor $\mathbf{S}$.

Solution

The first invariant is the trace of the tensor $\mathbf{S}$:

$$I_1 = tr\,(\mathbf{S}) = \sigma_{kk} = \sigma_{11} + \sigma_{22} + \sigma_{33} = 1 + 3 - 2 = 2$$

The second and third invariants can be calculated accordingly:

$$I_2 = \frac{1}{2}((tr\,\mathbf{S})^2 - tr\,(\mathbf{S} \cdot \mathbf{S})) = \frac{1}{2}(\sigma_{jj}\sigma_{kk} - \sigma_{jk}\sigma_{kj})$$

$$I_3 = \det(\mathbf{S}) = -6 + 8 - 3 = -1$$

By utilizing matrix multiplication, we obtain

$$\mathbf{S} \cdot \mathbf{S} = \begin{pmatrix} 6 & 8 & -1 \\ 8 & 13 & 2 \\ -1 & 2 & 5 \end{pmatrix} (\text{MPa})^2$$

So the trace of $\mathbf{S} \cdot \mathbf{S}$ is

$$tr\,(\mathbf{S} \cdot \mathbf{S}) = 6 + 13 + 5 = 24$$

Therefore, the second invariant I_2 is

$$I_2 = \frac{1}{2}((tr\,\mathbf{S})^2 - tr\,(\mathbf{S} \cdot \mathbf{S}))$$

$$I_2 = \frac{1}{2}[(2)^2 - (24)] = -10$$

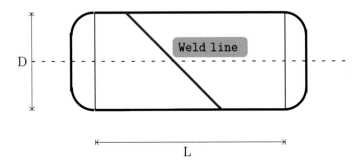

Fig. 6.3 Example 3.3.1

Now we substitute these invariants in the characteristic polynomial and obtain

$$-\lambda^3 + 2\lambda^2 + 10\lambda - 1 = 0$$

The principal values are obtained after solving the above polynomial, and they are

$$\lambda_1 = 4.98\,\text{MPa}, \lambda_2 = 0.1\,\text{MPa}, \lambda_3 = -2.38\,\text{MPa}$$

An alternative solution is to find the determinant expressed in Eq. (3.4). Here, we have

$$\det \begin{pmatrix} 1 - \lambda & 2 & 1 \\ 2 & 3 - \lambda & 0 \\ 1 & 0 & -2 - \lambda \end{pmatrix} = 0$$

After expansion and rearranging the component, we obtain

$$-\lambda^3 + 2\lambda^2 + 10\lambda - 1 = 0$$

the characteristic equation for the tensor **S**. Notice that the invariants are the coefficient of λ^0, λ^1 and λ^2, respectively.

Example 3.3.2

A pressure vessel shown in Fig. 3.3 is made of a cylinder welded by caps at both ends. The cap is made of a deep-drawn circular plate, while the cylinder is made of a rectangular plate rolled and welded at the end at the angle of 45° from the horizontal axis. The inner diameter of the cylinder is $D = 2R$, and the thickness of the plate is t. The vessel is filled with liquid of pressure p. Assuming that the vessel is a thin-walled vessel, i.e. $D/t > 10$, calculate the stresses in the vessel. Use your results to calculate the stress along the indicated weld beads or line (red line) (Fig. 6.3).

Numerical values: $p = 3\,\text{MPa}$, $D = 2\,\text{m}$, $L = 5\,\text{m}$ and $t = 10\,\text{mm}$.

Solution

With the assumption that the vessel is thin-walled and the loading from gas pressure is isotropic, the stresses in the vessel resulting from the loading are only in axial and circumferential directions. Stress in the radial direction is considered to be negligibly small. Let e_1 be the axial axis and e_2 be the circumferential axis in the wall. e_3 is the radial axis normal to the cylindrical wall. The force exerted by the liquid on the wall along its axial axis is $p(\pi R^2)$ and this force is balanced by the force resulting in the wall, which is $\sigma_{11}(2\pi R)$. Therefore, the traction or stress σ_{11} is

$$p(\pi R^2) = \sigma_{11}(2\pi R)t$$

or

$$\sigma_{11} = \frac{pR}{2t}$$

Similarly for the stress perpendicular to the wall, we have to balance the force exerted by the liquid to the wall and force resulting in the vessel's wall. That is

$$pRL = \sigma_{22}Lt$$

which produces

$$\sigma_{22} = \frac{pR}{t}$$

With no other stress inside the wall, the stress tensor can be written as

$$(\mathbf{S}) = \begin{pmatrix} pR/2t & 0 & 0 \\ 0 & pR/t & 0 \\ 0 & 0 & 0 \end{pmatrix}$$

In order to find the stress components in the weld beads, we need to do an orthogonal transformation of the stress tensor obtained previously. Basically, we rotate the current axis 45° clockwise to capture the weld line. The orthogonal or rotation tensor for this purpose is $\mathbf{Q}$,[2] which is

$$(\mathbf{Q}) = \begin{pmatrix} \cos(45) & \sin(45) & 0 \\ -\sin(45) & \cos(45) & 0 \\ 0 & 0 & 1 \end{pmatrix}$$

or

$$(\mathbf{Q}) = \begin{pmatrix} \sqrt{2}/2 & \sqrt{2}/2 & 0 \\ -\sqrt{2}/2 & \sqrt{2}/2 & 0 \\ 0 & 0 & 1 \end{pmatrix}$$

---

[2] Refer Appendix 1 for orthogonal transformation

After the orthogonal transformation

$$\tilde{S} = Q \cdot S \cdot Q^T$$

we obtain

$$(\tilde{S}) = \begin{pmatrix} 3pR/4t & pR/4t & 0 \\ pR/4t & 3pR/4t & 0 \\ 0 & 0 & 1 \end{pmatrix}$$

the same Cauchy stress tensor but written based on the rotated axis system. With the data provided, the stress components in the weld line are

$$\tilde{\sigma}_{11} = 3pR/4t = 450 \, \text{MPa}$$

$$\tilde{\sigma}_{22} = 3pR/4t = 450 \, \text{MPa}$$

$$\tilde{\sigma}_{11} = pR/4t = 150 \, \text{MPa}$$

Problem 3.3.1

Given is the Cauchy stress tensor **S** as follows:

$$(S) = \begin{pmatrix} x_1 + x_2 & f(x_1, x_2) & 0 \\ f(x_1, x_2) & x_1 - 2x_2 & 0 \\ 0 & 0 & x_2 \end{pmatrix}$$

1. Determine the function $f(x_1, x_2)$ assuming that the body force is negligibly small and the origin is stress-free.
2. Determine the stress vector at any point on the surface $x_1 = constant$ and $x_2 = constant$.

Problem 3.3.2

An observer $\mathcal{A}$ measures the Cauchy stress tensor at one particular point of a body and announces that his results are as expressed below

$$(S_A) = \begin{pmatrix} 10 & 5 & 18 \\ 5 & 10 & 3 \\ 18 & 3 & 15 \end{pmatrix}$$

The observer $\mathcal{B}$ measures the Cauchy stress tensor at the same point using different instrumentation and declares his result as

$$(S_B) = \frac{1}{9} \begin{pmatrix} 226 & 35 & -104 \\ 35 & -26 & 41 \\ -104 & 41 & 115 \end{pmatrix}$$

Do these observers measure the same state of stress at that point?

Solution

Two tensors are the same if their invariants are the same. The three invariants of the stress tensors S_A and S_B are

$$I_{S_A} = tr(S_A) = 10 + 10 + 15 = 35$$

$$II_{S_A} = \det \begin{pmatrix} 10 & 5 \\ 5 & 10 \end{pmatrix} + \det \begin{pmatrix} 10 & 3 \\ 3 & 15 \end{pmatrix} + \det \begin{pmatrix} 10 & 18 \\ 18 & 15 \end{pmatrix} = 42$$

$$III_{S_A} = \det \begin{pmatrix} 226 & 35 & -104 \\ 35 & -26 & 41 \\ -104/ & 41 & 115 \end{pmatrix} = -1665$$

The invariants of the stress tensor S_B are calculated in a similar manner.

$$I_{S_B} = tr(S_B) = (1/9)(226 - 26 + 115) = 35$$

$$II_{S_B} = \det(1/9) \begin{pmatrix} 226 & 35 \\ 35 & -26 \end{pmatrix} + \det(1/9) \begin{pmatrix} -26 & 41 \\ 41 & 115 \end{pmatrix}$$

$$+ \det(1/9) \begin{pmatrix} 226 & -104 \\ -104 & 115 \end{pmatrix} = 42$$

$$III_{S_B} = \det(1/9) \begin{pmatrix} 226 & 35 & -104 \\ 35 & -26 & 41 \\ -104 & 41 & 115 \end{pmatrix} = -1665$$

Problem 3.3.3

Determine the principal stresses and principal directions of the following Cauchy stress tensor:

$$(S) = \begin{pmatrix} 5 & 10 & 8 \\ 10 & 7 & -6 \\ 8 & -6 & 12 \end{pmatrix}$$

Perform an orthogonal transformation of the above tensor using the following orthogonal matrix:

$$(Q) = \frac{1}{3}\begin{pmatrix} 1 & 2 & 2 \\ 2 & 1 & -2 \\ -2 & 2 & -1 \end{pmatrix}$$

and find the principal stresses and principal directions of this transformed tensor. Based on your answer, what conclusion can you make?

Solution

The eigenvalues of S are

$$\lambda_1 = 17.3, \lambda_2 = 15.5, \lambda_3 = -8.8$$

The corresponding normalized eigenvectors are

$$n_1 = (0.63, 0.17, 0.76)^T, n_2 = (0.37, 0.79, -0.49)^T, n_3 = (0.68, -0.59, -0.43)^T$$

The orthogonal transformation of the stress tensor S produces

$$\hat{S} = Q \cdot S \cdot Q^T$$

$$(\hat{S}) = \begin{pmatrix} 35/3 & 6 & -26/3 \\ 6 & 25/3 & 28/3 \\ -26/3 & 28/3 & 4 \end{pmatrix}$$

This transformed stress tensor has the following eigenvalues

$$\hat{\lambda}_1 = 17.3115, \hat{\lambda}_2 = 15.509, \hat{\lambda}_3 = -8.820$$

which indicate that the tensors S and $\hat{S}$ are the same stress tensors.

Problem 3.3.4

Show that the symmetrical Cauchy stress tensor S has a set of real eigenvalues and if, furthermore, they are distinct, the associated principal directions form an orthonormal basis system.

Solution

Firstly, we show that the Cauchy stress tensor possesses three real eigenvalues $\lambda_a, \lambda_b, \lambda_c$. If λ is the eigenvalue of $\mathbf{S}$, then the characteristic equation for λ is given as follows:

$$(\lambda - a)(\lambda - b)(\lambda - c) = 0$$

whereby a, b and c are the eigenvalues of the symmetrical tensor $\mathbf{S}$, which are assumed to be complex numbers. After expanding the above equation, we have

$$\lambda^3 - (a + b + c)\lambda^2 + (ab + bc + ac)\lambda - abc = 0$$

By comparing with the characteristic equation of real symmetrical tensor $\mathbf{S}$ in Chap. 3, Sect. 3.3, we have

$$-\lambda^3 + I_1\lambda^2 - I_2\lambda + I_3 = 0$$

with

$$I_1 = tr\,(\mathbf{S}) = \sigma_{kk}$$

$$I_2 = \frac{1}{2}((tr\,\mathbf{S})^2 - tr\,(\mathbf{S} \cdot \mathbf{S})) = \frac{1}{2}(\sigma_{jj}\sigma_{kk} - \sigma_{jk}\sigma_{kj})$$

$$I_3 = \det(\mathbf{S})$$

$$(a + b + c) = tr\,(\mathbf{S}) = \sigma_{11} + \sigma_{22} + \sigma_{33}$$

]
$$(ab + bc + ac) = 1/2[(\sigma_{11}^2 + \sigma_{22}^2 + \sigma_{33}^2) - \sigma_{12}\sigma_{21} - \sigma_{13}\sigma_{31} - \sigma_{32}\sigma_{23}]$$

and

$$abc = \det(\mathbf{S})$$

The values on the right side of the above three equations are real numbers. This in fact contradicts the previous assumption that a, b and c are complex numbers. Therefore, we can conclude that a, b and c are real eigenvalues of the Cauchy stress tensor $\mathbf{S}$.

Secondly, we assume that the symmetrical Cauchy stress tensor $\mathbf{S}$ has three distinctive real eigenvalues, $\lambda_a \neq \lambda_b \neq \lambda_c$ and three associated eigenvectors $\mathbf{a}, \mathbf{b}$ and $\mathbf{c}$. Then we can write

$$\mathbf{S} \cdot \mathbf{a} = \lambda_a \mathbf{a}$$

$$\mathbf{S} \cdot \mathbf{b} = \lambda_b \mathbf{b}$$

$$\mathbf{S} \cdot \mathbf{c} = \lambda_c \mathbf{c}$$

Now we perform a scalar product of the first equation with the vector $\mathbf{b}$ and the second equation with the vector $\mathbf{a}$ and obtain

$$\mathbf{b} \cdot \mathbf{S} \cdot \mathbf{a} = \lambda_a \mathbf{b} \cdot \mathbf{a}$$

and

$$\mathbf{a} \cdot \mathbf{S} \cdot \mathbf{b} = \lambda_b \mathbf{a} \cdot \mathbf{b}$$

By subtracting both equations from each other, we have

$$\mathbf{a} \cdot \mathbf{S} \cdot \mathbf{b} - \mathbf{b} \cdot \mathbf{S} \cdot \mathbf{a} = \lambda_b \mathbf{a} \cdot \mathbf{b} - \lambda_a \mathbf{b} \cdot \mathbf{a}$$

Since $\mathbf{S}$ is symmetrical and the scalar product $\mathbf{a} \cdot \mathbf{b}$ is commutative, we can rewrite the above equation as

$$0 = (\lambda_b - \lambda_a)\mathbf{a} \cdot \mathbf{b}$$

Since $\lambda_b \neq \lambda_a$, the above equation is only valid if and only if $\mathbf{a} \cdot \mathbf{b} = 0$, which implies that these two vectors (eigenvectors) are orthogonal to each other. This line of proof can be used for other pairings, $(\mathbf{b}, \mathbf{c})$ and $(\mathbf{c}, \mathbf{a})$. Q.e.d.

Problem 3.3.6

Figure 3.4 illustrates a rectangular bar under the axial loading and the components of the resulting first Piola-Kirchhoff stress tensor is given below

$$(\mathbf{T}) = \begin{pmatrix} \sigma & 0 & 0 \\ 0 & 0 & 0 \\ 0 & 0 & 0 \end{pmatrix}$$

Calculate the components of the stress tensor after the body being stretched by $\alpha\%$ of its original length in axial direction x to the current configuration, and rotation about x_3 axis, as shown in Fig. 3.5.

Solution

In this problem, the body undergoes two deformation processes, stretching and rotation. The deformation gradient $\mathbf{F}$ constitutes therefore a stretch tensor $\mathbf{U}$ and an orthogonal rotation tensor $\mathbf{R}$. According to the polar decomposition theorem,

$$\mathbf{F} = \mathbf{R} \cdot \mathbf{U}$$

since the body is stretched first and then rotated about x_3 axis anticlockwise, as seen from the x_3 axis. The left stretch tensor $\mathbf{U}$ can be written as

$$\mathbf{U} = \begin{pmatrix} \alpha & 0 & 0 \\ 0 & \beta & 0 \\ 0 & 0 & \gamma \end{pmatrix}$$

where the β and γ are resulting stretching in the other two axes. In a uniaxial stretching, $\beta = \gamma = 1/\sqrt{\alpha}$:

$$\mathbf{U} = \begin{pmatrix} \alpha & 0 & 0 \\ 0 & 1/\sqrt{\alpha} & 0 \\ 0 & 0 & 1/\sqrt{\alpha} \end{pmatrix}$$

The body is then rotated by 90° about x_3 axis, therefore the rotation tensor can be obtained as

$$\mathbf{R} = \begin{pmatrix} 0 & 1 & 0 \\ -1 & 0 & 0 \\ 0 & 0 & 1 \end{pmatrix}$$

which produces the deformation gradient $\mathbf{F}$ as follows:

$$(\mathbf{F}) = \begin{pmatrix} 0 & 1 & 0 \\ -1 & 0 & 0 \\ 0 & 0 & 1 \end{pmatrix} \begin{pmatrix} \alpha & 0 & 0 \\ 0 & 1/\sqrt{\alpha} & 0 \\ 0 & 0 & 1/\sqrt{\alpha} \end{pmatrix}$$

$$(\mathbf{F}) = \begin{pmatrix} 0 & 1/\sqrt{\alpha} & 0 \\ -\alpha & 0 & 0 \\ 0 & 0 & 1/\sqrt{a} \end{pmatrix}$$

There are two situations that arise here: 1. Finding the Cauchy stress tensor, i.e. tensor which describes the state of stress of a body based on the current configuration. 2. Finding the second Piola- Kirchhoff tensor associated with the given first Piola-Kirchhoff tensor.

1. Finding the Cauchy stress tensor. The relationship between the Cauchy stress tensor and the first Piola-Kirchhoff tensor is given as follows:

$$\mathbf{S} = J\mathbf{F} \cdot \mathbf{T}$$

$$(\mathbf{S}) = \begin{pmatrix} 0 & 1/\sqrt{\alpha} & 0 \\ -\alpha & 0 & 0 \\ 0 & 0 & 1/\sqrt{a} \end{pmatrix} \begin{pmatrix} \sigma & 0 & 0 \\ 0 & 0 & 0 \\ 0 & 0 & 0 \end{pmatrix}$$

$$(\mathbf{S}) = \begin{pmatrix} 0 & 0 & 0 \\ -\alpha\sigma & 0 & 0 \\ 0 & 0 & 0 \end{pmatrix}$$

With $J = 1$. 2. For finding the second Piola-Kirchhoff tensor $\hat{\mathbf{T}}$, we use the relation below (Figs. 6.4 and 6.5)

Fig. 6.4 Initial configuration

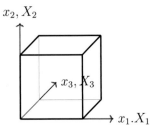

Fig. 6.5 After stretching

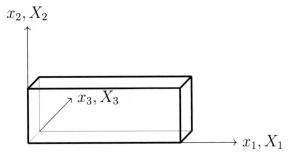

$$\hat{\mathbf{T}}^T = \mathbf{F}^{-1} \cdot \mathbf{T}^T$$

$$(\hat{\mathbf{T}}^T) = \begin{pmatrix} 0 & -1/\alpha & 0 \\ \sqrt{\alpha} & 0 & 0 \\ 0 & 0 & \sqrt{\alpha} \end{pmatrix} \begin{pmatrix} \sigma & 0 & 0 \\ 0 & 0 & 0 \\ 0 & 0 & 0 \end{pmatrix}$$

$$(\hat{\mathbf{T}}^T) = \begin{pmatrix} 0 & 0 & 0 \\ \sqrt{\alpha}\sigma & 0 & 0 \\ 0 & 0 & 0 \end{pmatrix}$$

6.12 Section 3.4

Example 3.4.1

Given are the deformation gradient $\mathbf{F}$ and the Cauchy stress tensor as follows:

$$(\mathbf{F}) = \begin{pmatrix} 0.935 & 0.312 & 0.231 \\ -0.210 & 1.12 & 0.11 \\ -0.32 & 0.11 & 1.12 \end{pmatrix}$$

$$(S) = \begin{pmatrix} 5 & 10 & 8 \\ 10 & 7 & -6 \\ 8 & -6 & 12 \end{pmatrix}$$

Calculate the first Piola-Kirchhoff tensor and corresponding second Piola-Kirchhoff stress tensors.

Solution

The Jacobian J is calculated from the determinant of the deformation gradient

$$J = \det(\mathbf{F}) = \det \begin{pmatrix} 0.935 & 0.312 & 0.231 \\ -0.210 & 1.12 & 0.11 \\ -0.32 & 0.11 & 1.12 \end{pmatrix} = 1.3014$$

The inverse of the deformation gradient $\mathbf{F}$ is readily calculated using MS Excel and obtained as

$$(\mathbf{F}^{-1}) = \begin{pmatrix} 0.955 & -0.249 & -0.172 \\ 0.154 & 0.861 & -0.116 \\ 0.258 & -0.156 & 0.855 \end{pmatrix}$$

Now as defined in Eq. (3.8), the first Piola-Kirchhoff tensor is

$$\mathbf{T} = (1/J)\mathbf{F}^{-1} \cdot \mathbf{S}$$

$$\mathbf{T} = (1/1.301) \begin{pmatrix} 0.955 & -0.249 & -0.172 \\ 0.154 & 0.861 & -0.116 \\ 0.258 & -0.156 & 0.855 \end{pmatrix} \cdot \begin{pmatrix} 5 & 10 & 8 \\ 10 & 7 & -6 \\ 8 & -6 & 12 \end{pmatrix}$$

$$= \begin{pmatrix} 0.291 & 2.847 & 2.275 \\ 2.723 & 2.663 & -1.719 \\ 2.117 & -1.174 & 4.271 \end{pmatrix}$$

Note that the first Piola-Kirchhoff is not symmetrical. The second Piola-Kirchhoff tensor can be obtained from Eq. (3.10):

$$\tilde{\mathbf{T}}^{T} = (1/J)\mathbf{F}^{-1} \cdot \mathbf{S} \cdot \mathbf{F}^{-T}$$

Basically

$$\tilde{\mathbf{T}}^{T} = \begin{pmatrix} 0.291 & 2.847 & 2.275 \\ 2.723 & 2.663 & -1.719 \\ 2.117 & -1.174 & 4.271 \end{pmatrix} \cdot \begin{pmatrix} 0.935 & 0.312 & 0.231 \\ -0.210 & 1.12 & 0.11 \\ -0.32 & 0.11 & 1.12 \end{pmatrix}^{-T}$$

$$\tilde{\mathbf{T}}^T = \begin{pmatrix} -0.823 & 2.233 & 1.577 \\ 2.233 & 2.912 & -1.183 \\ 1.577 & -1.183 & 4.380 \end{pmatrix}$$

Unlike the first Piola-Kirchhoff tensor, the second Piola-Kirchhoff tensor is symmetrical.

Problem 3.4.1

Based on the Cauchy stress tensor and the deformation gradient given in **Example 3.4.1** above, calculate the principal stresses and principal directions of the second Piola-Kirchhoff tensor and compare the results with the principal stresses and principal directions of the Cauchy stress tensor.

Solution

Problem 3.4.2

The relation between the Cauchy and the second Piola-Kirchhoff stress tensor is given in Eq. (3.10) or (3.11):

$$\mathbf{S} = (1/J)\,\mathbf{F} \cdot \tilde{\mathbf{T}} \cdot \mathbf{F}^T$$

Construct a tensor of fourth-order $\underline{\mathbf{F}}$ so that the above relation can be written as

$$\mathbf{S} = \underline{\mathbf{F}} \cdot \cdot \tilde{\mathbf{T}}$$

with "··" being a double scalar product between tensors of higher order. Find also its inverse $\underline{\mathbf{F}}^{-1}$.

Solution

From the equation
$$\mathbf{S} = (1/J)\,\mathbf{F} \cdot \tilde{\mathbf{T}} \cdot \mathbf{F}^T$$

and
$$\mathbf{S} = \underline{\mathbf{F}} \cdot \cdot \tilde{\mathbf{T}}$$

It is obvious that the fourth-order tensor $\underline{\mathbf{F}}$ is made up of a dyadic product between two deformation gradient tensors of second-order, i.e. $\underline{\mathbf{F}} \sim \mathbf{F} \otimes \mathbf{F}$, so that

$$\underline{\mathbf{F}} \cdot \cdot \tilde{\mathbf{T}} = (1/J)\,\mathbf{F} \cdot \tilde{\mathbf{T}} \cdot \mathbf{F}^T \quad \longleftrightarrow \quad (*)$$

In indicial notation, $\underline{\mathbf{F}}$ can be written as

$$\underline{\mathbf{F}} = F_{ijkl}\,\mathbf{e}_{ijkl}$$

with $\mathbf{e}_{ijkl}$ being a dyadic product of four basis vectors $\mathbf{e}_i$

$$\mathbf{e}_{ijkl} = \mathbf{e}_i \otimes \mathbf{e}_j \otimes \mathbf{e}_k \otimes \mathbf{e}_l$$

Since $\underline{\mathbf{F}}$ is made up of a dyadic product between two deformation gradient tensors F_{kl}, for example,

$$F_{ijkl}\mathbf{e}_{ijkl} = (1/J)\,F_{kl}F_{ji}\,\mathbf{e}_{ijkl}$$

However, the above index permutation of two deformation gradients $F_{kl}F_{ji}$ does not fulfil the definition equation (*) for the fourth-order tensor $\underline{\mathbf{F}}$ above. The indices for F have to be rearranged so that they satisfy the equation (*). Out of 24 permutations of indices, there are two distinct index permutations which produce the definition equation (*). These permutations are $F_{ik}F_{jl}$ and $F_{il}F_{jk}$. Thus, the fourth-order tensor $\underline{\mathbf{F}}$ can be constructed as follows:

$$F_{ijkl}\,\mathbf{e}_{ijkl} = (1/J)\,(1/2)[F_{ik}F_{jl} + F_{il}F_{jk}]\,\mathbf{e}_{ijkl}$$

$$= \frac{1}{2J}\,(F_{ik}F_{jl} + F_{il}F_{jk})\,\mathbf{e}_{ijkl}$$

The fourth-order deformation gradient tensor, in indicial notation, is

$$F_{ijkl} = \frac{1}{2J}\,(F_{ik}F_{jl} + F_{il}F_{jk})$$

To verify the above tensor, we perform the double scalar product between $\underline{\mathbf{F}}$ and the second Piola-Kirchhoff tensor $\tilde{\mathbf{T}}$ in indicial notation. From

$$\mathbf{S} = \underline{\mathbf{F}} \cdot\cdot\, \tilde{\mathbf{T}}$$

$$= F_{ijkl}\,\mathbf{e}_{ijkl} \cdot\cdot\, \tilde{T}_{mn}\mathbf{e}_{mn}$$

$$= \frac{1}{2J}\,(F_{ik}F_{jl} + F_{il}F_{jk})\,\mathbf{e}_{ijkl} \cdot\cdot\, \tilde{T}_{mn}\mathbf{e}_{mn}$$

$$= \frac{1}{2J}\,(F_{ik}F_{jl} + F_{il}F_{jk})\,\tilde{T}_{mn}\delta_{lm}\delta_{kn}\mathbf{e}_{ij}$$

$$= \frac{1}{2J}\,(F_{ik}F_{jl} + F_{il}F_{jk})\,\tilde{T}_{lk}\mathbf{e}_{ij}$$

$$= \frac{1}{2J} (F_{ik} F_{jl} \tilde{T}_{lk} + F_{il} F_{jk} \tilde{T}_{lk}) \mathbf{e}_{ij}$$

$$= \frac{1}{2J} (F_{jl} \tilde{T}_{lk} F_{ik} + F_{il} \tilde{T}_{lk} F_{jk}) \mathbf{e}_{ij}$$

$$= \frac{1}{2J} (\mathbf{F} \cdot \tilde{\mathbf{T}} \cdot \mathbf{F}^T + \mathbf{F} \cdot \tilde{\mathbf{T}} \cdot \mathbf{F}^T)$$

Therefore, we have

$$\mathbf{S} = \underline{\mathbf{F}} \cdot \cdot \tilde{\mathbf{T}} = \frac{1}{2J} \mathbf{F} \cdot \tilde{\mathbf{T}} \cdot \mathbf{F}^T, \quad q.e.d$$

The same method can be used to construct an inverse of the fourth-order tensor $\underline{\mathbf{F}}$. Here, we start with the equation

$$\tilde{\mathbf{T}} = J \mathbf{F}^{-1} \cdot \mathbf{S} \cdot \mathbf{F}^{-T}$$

$$\tilde{\mathbf{T}} = \underline{\mathbf{G}} \cdot \cdot \mathbf{S}$$

The fourth-order tensor $\underline{\mathbf{G}}$ is the inverse of $\underline{\mathbf{F}}$. Taking $\underline{\mathbf{G}} = G_{ijkl} \mathbf{e}_{ijkl}$ and $\mathbf{G} = \mathbf{F}^{-1}$, we can arrange the index of the fourth-order tensor $\underline{\mathbf{G}}$ using the proper index permutation as discussed before. Here, we have

$$G_{ijkl} \mathbf{e}_{ijkl} = J (1/2)[G_{ik} G_{jl} + G_{il} G_{jk}] \mathbf{e}_{ijkl}$$

In terms of the inverse of the deformation gradient tensor, $\mathbf{G} = \mathbf{F}^{-1}$ the fourth-order tensor $\underline{\mathbf{G}}$ takes the following form:

$$G_{ijkl} \mathbf{e}_{ijkl} = \frac{J}{2} [F_{ik}^{-1} F_{jl}^{-1} + F_{il}^{-1} F_{jk}^{-1}] \mathbf{e}_{ijkl}$$

Problem 3.4.3

Given is a body with the motion

$$\begin{aligned}
x_1 &= X_1 + (1/4)X_2^2 \\
x_2 &= (X_1 + 1)X_2^2 \\
x_3 &= X_3
\end{aligned} \tag{6.2}$$

The resulting Cauchy stress tensor $\mathbf{S}(X_i)$ as a function of material points $X_i, i = 1, 2, 3$ is determined as follows:

$$(S) = \begin{pmatrix} X_1^2 & X_1X_2 & 0 \\ X_1X_2 & X_2^2 & X_2X_3^2 \\ 0 & X_2X_3^2 & X_2^2 \end{pmatrix}$$

Show that the Eq. (3.12) represent a motion and calculate the first and second Piola-Kirchhoff tensors at the material point P (1, 0.5, 1.5).

Solution

The first step is to obtain the deformation gradient **F**

$$(F) = \begin{pmatrix} 1 & (1/2)X_2 & 0 \\ X_2^2 & 2(X_1 + 1)X_2 & 0 \\ 2 & 0 & (1/2)X_3 \end{pmatrix}$$

At the material point P, the deformation gradient **F** takes the following form:

$$(F(P)) = \begin{pmatrix} 1 & 0.25 & 0 \\ 0.25 & 2. & 0 \\ 2 & 0 & 0.75 \end{pmatrix}$$

and the Cauchy stress tensor **S** at the point P

$$(S(P)) = \begin{pmatrix} 1 & 0.5 & 0 \\ 0.5 & 0.25 & 1.125 \\ 0 & 1.125 & 2.25 \end{pmatrix}$$

For obtaining the first Piola-Kirchhoff stress tensor, one can use the following equation:

$$S = JF \cdot T$$

or

$$T = (1/J)F^{-1} \cdot S$$

$$(T) = (1/1.45) \begin{pmatrix} 1 & 0.25 & 0 \\ 0.25 & 2. & 0 \\ 2 & 0 & 0.75 \end{pmatrix}^{-1} \begin{pmatrix} 1 & 0.5 & 0 \\ 0.5 & 0.25 & 1.125 \\ 0 & 1.125 & 2.25 \end{pmatrix}$$

which delivers the first Piola-Kirchhoff stress tensor

$$(T(P)) = \begin{pmatrix} 0.667 & 0.334 & -0.1 \\ 0.089 & 0.044 & 0.4 \\ -1.78 & 0.145 & 2.336 \end{pmatrix}$$

Unlike the Cauchy stress tensor, the first Piola-Kirchhoff tensor is asymmetrical. The symmetrization of this tensor produces the second Piola-Kirchhoff stress tensor $\hat{\mathbf{T}}$:

$$\hat{\mathbf{T}}^T = \mathbf{F}^{-1} \cdot \mathbf{T}^T$$

$$(\hat{\mathbf{T}}^T) = \begin{pmatrix} 1 & 0.25 & 0 \\ 0.25 & 2. & 0 \\ 2 & 0 & 0.75 \end{pmatrix}^{-1} \begin{pmatrix} 0.667 & 0.334 & -0.1 \\ 0.089 & 0.044 & 0.4 \\ -1.78 & 0.145 & 2.336 \end{pmatrix}^T$$

We have then the second Piola-Kirchhoff stress tensor:

$$\hat{\mathbf{T}}^T (P)) = \begin{pmatrix} 0.645879 & 0.0861172 & -1.85583 \\ 0.0861172 & 0.0114823 & 0.304281 \\ -1.85583 & 0.304281 & 8.06344 \end{pmatrix}$$

Based on this example, one can see clearly the symmetry of the second Piola-Kirchhoff.

6.13 Section 3.5

Example 3.5.1

Show that the net-stress tensor

$$\hat{\mathbf{S}} = \Psi^{-1} \cdot \mathbf{S}$$

is not symmetric and derive the symmetrical and asymmetrical parts of the net-stress tensor $\hat{\mathbf{S}}$.

Solution

The Cauchy stress tensor $\mathbf{S}$ and the continuity tensor Ψ are symmetrical. From the equation

$$\mathbf{S} = \Psi \cdot \hat{\mathbf{S}}$$

we get

$$\mathbf{S} = \Psi \cdot \hat{\mathbf{S}} = \mathbf{S}^T = (\Psi \cdot \hat{\mathbf{S}})^T = \hat{\mathbf{S}}^T \cdot \Psi^T$$

or

$$\Psi \cdot \hat{\mathbf{S}} = \hat{\mathbf{S}}^T \cdot \Psi$$

$$\hat{\mathbf{S}} = \Psi^{-1} \cdot \hat{\mathbf{S}}^T \cdot \Psi$$

which shows that the net-stress tensor $\hat{\mathbf{S}}$ is not symmetrical. The transposition of the net-stress tensor $\hat{\mathbf{S}}$ is

$$\hat{\mathbf{S}}^T = (\Psi^{-1} \cdot \hat{\mathbf{S}}^T \cdot \Psi)^T = \Psi^T \cdot \hat{\mathbf{S}} \cdot \Psi^{-T} = \Psi \cdot \hat{\mathbf{S}} \cdot \Psi^{-1}$$

The symmetrical part of the net-stress tensor is

$$Sym\,\hat{\mathbf{S}} = \frac{1}{2}(\Psi^{-1} \cdot \hat{\mathbf{S}}^T \cdot \Psi + \Psi \cdot \hat{\mathbf{S}} \cdot \Psi^{-1})$$

and its asymmetrical counterpart is

$$Assym\,\hat{\mathbf{S}} = \frac{1}{2}(\Psi^{-1} \cdot \hat{\mathbf{S}}^T \cdot \Psi - \Psi \cdot \hat{\mathbf{S}} \cdot \Psi^{-1})$$

Problem 3.5.1

Derive the equation of motion of a body with a damaged continuum, assuming the body force is negligibly small. The damage tensor is given as

$$\Omega = \begin{pmatrix} 1 - \alpha & 0 & 0 \\ 0 & 1 - \beta & 0 \\ 0 & 0 & 1 - \gamma \end{pmatrix}$$

whereby the damage parameters $1 - \alpha$, $1 - \beta$ and $1 - \gamma$ take the values between 0 and 1.

Solution

By neglecting the body forces, the equation of motion of a body is given as follows:

$$\frac{\partial(\rho\mathbf{v})}{\partial t} = \nabla \cdot \mathbf{S}$$

Replacing the Cauchy stress tensor $\mathbf{S}$ with the net-stress tensor of a damaged continuum, we have the following equation of motion:

$$\frac{\partial(\rho\mathbf{v})}{\partial t} = \nabla \cdot (\Psi \cdot \hat{\mathbf{S}})$$

or

$$\frac{\partial(\rho\mathbf{v})}{\partial t} = \nabla \cdot ((\mathbf{I} - \Omega) \cdot \hat{\mathbf{S}})$$

Since the damage tensor is independent of position, i.e. is constant everywhere, the above equation can be simplified as follows:

$$\frac{\partial(\rho\mathbf{v})}{\partial t} = (\mathbf{I} - \mathbf{\Omega}) \cdot \nabla \cdot \hat{\mathbf{S}}$$

Rewriting the above equation in the indicial form, we have

$$\frac{\partial(\rho v_i)}{\partial t} = (\delta_{ik} - \Omega_{ik})\,\hat{\sigma}_{kj,j}$$

After expansion, for example, we have

$$\frac{\partial(\rho v_1)}{\partial t} = (1 - \alpha)[\frac{\partial(\hat{\sigma}_{11})}{\partial x_1} + \frac{\partial(\hat{\sigma}_{12})}{\partial x_2} + \frac{\partial(\hat{\sigma}_{13})}{\partial x_3}]$$

$$\frac{\partial(\rho v_2)}{\partial t} = (1 - \beta)[\frac{\partial(\hat{\sigma}_{21})}{\partial x_1} + \frac{\partial(\hat{\sigma}_{22})}{\partial x_2} + \frac{\partial(\hat{\sigma}_{23})}{\partial x_3}]$$

$$\frac{\partial(\rho v_3)}{\partial t} = (1 - \gamma)[\frac{\partial(\hat{\sigma}_{31})}{\partial x_1} + \frac{\partial(\hat{\sigma}_{32})}{\partial x_2} + \frac{\partial(\hat{\sigma}_{33})}{\partial x_3}]$$

From these equations, we note that if the case of quasi-static deformation the stress fields of damaged and undamaged continuum are almost similar.

6.14 Section 4.1

Example 4.4.1

Find the inversion of the material equation (4.12).

Solution

The double scalar product of the equation (4.12) delivers the trace of the linearized Lagrangian strain tensor $\hat{\mathbf{E}}$,

$$\mathbf{I} \cdot \cdot \mathbf{S} = tr(\mathbf{S}) = \mathbf{I} \cdot \cdot \lambda\, tr(\hat{\mathbf{E}})\mathbf{I} + 2\mu \mathbf{I} \cdot \cdot \hat{\mathbf{E}}$$

With $\mathbf{I} \cdot \cdot \mathbf{I} = 3$, the above equation reduces to

$$\mathbf{I} \cdot \cdot \mathbf{S} = tr(\mathbf{S}) = (3\lambda + 2\mu)\, tr(\hat{\mathbf{E}})$$

From Eq. (4.12), we have then

$$\hat{\mathbf{E}} = \frac{1}{2\mu}(\mathbf{S} - \lambda\, tr(\hat{\mathbf{E}})\mathbf{I}) = \frac{1}{2\mu}\left(\mathbf{S} - \frac{\lambda}{3\lambda + 2\mu}tr(\mathbf{S})\mathbf{I}\right)$$

Problem 4.1.1

Solution

6.15 Section 4.2

Example 4.2.1

Verify the relationship between the Lamè constants and the experimentally determined elastic and bulk moduli, E, G and the Poisson ratio ν as described by equations (6.3).

Solution

We can start with the Lame–Navier equation (4.12) and use it for a simple one-dimensional extension test. That is

$$\sigma_{pq} = \lambda\delta_{pq}\epsilon_{ss} + 2\mu(\epsilon_{pq} + \epsilon_{qp}) \longleftrightarrow \mathbf{S} = \lambda\, tr(\hat{\mathbf{E}})\mathbf{I} + 2\mu\hat{\mathbf{E}}$$

and the Cauchy stress tensor for a simple one-dimensional extension is given as follows:

$$(\mathbf{S}) = \begin{pmatrix} \sigma_{11} & 0 & 0 \\ 0 & 0 & 0 \\ 0 & 0 & 0 \end{pmatrix}$$

The linearized Lagrangian strain tensor for the extension test takes the following form:

$$(\hat{\mathbf{E}}) = \begin{pmatrix} \epsilon_{11} & 0 & 0 \\ 0 & \epsilon_{22} & 0 \\ 0 & 0 & \epsilon_{33} \end{pmatrix}$$

Let $\sigma_{11} = P$ be the stress along the extension axis, then $\epsilon_{11} = \frac{P}{E}$ and $\epsilon_{22} = \epsilon_{33} = -\frac{\nu P}{E}$, where E and ν are Young's modulus and Poisson's ratio, respectively. Therefore, the strain tensor now is

$$(\hat{\mathbf{E}}) = \begin{pmatrix} \frac{P}{E} & 0 & 0 \\ 0 & -\frac{\nu P}{E} & 0 \\ 0 & 0 & -\frac{\nu P}{E} \end{pmatrix}$$

Substituting these two tensors into the Lame–Navier Equation (4.12), we have

$$\begin{pmatrix} P\ 0\ 0 \\ 0\ 0\ 0 \\ 0\ 0\ 0 \end{pmatrix} = \lambda(\frac{P}{E} - 2\frac{\nu P}{E}) \begin{pmatrix} 1\ 0\ 0 \\ 0\ 1\ 0 \\ 0\ 0\ 1 \end{pmatrix} + 2\mu \begin{pmatrix} \frac{P}{E} & 0 & 0 \\ 0 & -\frac{\nu P}{E} & 0 \\ 0 & 0 & -\frac{\nu P}{E} \end{pmatrix}$$

and after simplification

$$P = \lambda(\frac{P}{E} - 2\frac{\nu P}{E}) + 2\frac{\nu P}{E}$$

and

$$0 = \lambda(\frac{P}{E} - 2\frac{\nu P}{E}) - 2\mu\frac{\nu P}{E}$$

By eliminating P, we obtain the relationship between the Lame constants and the elasticity constant E and ν:

$$\lambda = \frac{\nu E}{(1 - \nu)(1 - 2\nu)}$$

$$\mu = \frac{E}{2(1 + \nu)} = G$$

Example 4.2.2

The deformation gradient of a simple shear test is given as follows:

$$(\mathbf{F}) = \begin{pmatrix} 1\ \gamma\ 0 \\ 0\ 1\ 0 \\ 0\ 0\ 1 \end{pmatrix}$$

γ is the shear strain and in terms of stretch ratio $\lambda = l/l_0$, it takes the following form:

$$\gamma = \lambda - 1/\lambda$$

furthermore

$$\lambda_1 = \lambda, \lambda_2 = 1/\lambda, \lambda_3 = 1$$

Plot the Cauchy stress tensor versus stretch ratio if the sample for the testing used is a rubber block modelled as an incompressible Mooney–Rivlin material with $c_{10} = c_1 = 0.220\,\text{MPa}$ and $c_{01} = c_2 = 0.017\,\text{MPa}$. Determine the Cauchy stress if the sample is stretched to 30 % of its initial length.

Solution

From the given deformation gradient of a simple shear

$$(\mathbf{F}) = \begin{pmatrix} 1 & \gamma & 0 \\ 0 & 1 & 0 \\ 0 & 0 & 1 \end{pmatrix}$$

We determine firstly the right Cauchy–Green deformation tensor $\mathbf{B}$, together with its inverse $\mathbf{B}^{-1}$ to get the Cauchy stress tensor according to Eq. (4.15). After some linear algebraic manipulations, we have the corresponding right Cauchy–Green deformation tensor as follows:

$$(\mathbf{B}) = (\mathbf{F} \cdot \mathbf{F}^{\mathbf{T}}) = \begin{pmatrix} 1+\gamma^2 & \gamma & 0 \\ \gamma & 1 & 0 \\ 0 & 0 & 1 \end{pmatrix}$$

and its inverse $\mathbf{B}^{-1}$

$$(\mathbf{B}^{-1}) = \begin{pmatrix} 1 & -\gamma & 0 \\ -\gamma & 1+\gamma^2 & 0 \\ 0 & 0 & 1 \end{pmatrix}$$

Putting all these tensors together in Eq. (4.15), we have the Cauchy stress tensor

$$(\mathbf{S}) = \begin{pmatrix} -p+(c_1-c_2)+2c_2\gamma^2 & 2(c_1+c_2)\gamma & 0 \\ 2(c_1+c_2)\gamma & -p+(c_1-c_2)-2c_2\gamma^2 & 0 \\ 0 & 0 & -p+(c_1-c_2) \end{pmatrix}$$

and the pressure p is given as

$$p = \frac{2}{3}(c_1 I_1 - c_2 I_2)$$

$$= \frac{2}{3}[c_1(3+\gamma^2) - c_2(3+\gamma^2)]$$

$$= \frac{2}{3}(c_1 - c_2)(3+\gamma^2)$$

With $c_1 = 0.220$ MPa and $c_2 = 0.017$ MPa the normal stress, for instance σ_{11} is

$$\sigma_{11} = -\frac{2}{3}(c_1 - c_2)(3+\gamma^2) + (c_1 - c_2) + 2c_2\gamma^2$$

After simplification and substitution, we have

$$\sigma_{11} = -0.203 - 0.101\gamma^2$$

Since γ^2 is always positive, the normal stress for simple shear is always negative, thus indicating compression. For the case of stretch ratio equal to 30%, $\gamma = 30 - 1/30 = 29.97$. Therefore, the compression stress is 90.91 MPa.

Problem 4.2.1

A uniaxial tension test has been carried out on a sample of rubber with varying stretch ratios λ up to 2.5. The rubber sample is assumed to behave as a Mooney–Rivlin material with the strain energy function

$$\psi = c_1(I_{\mathbf{B}} - 3) + c_2(II_{\mathbf{B}} - 3)$$

and the Cauchy stress tensor

$$\mathbf{S} = \frac{\partial \psi}{\partial \mathbf{B}} = -p\mathbf{I} + 2c_1\mathbf{B} - 2c_2\mathbf{B}^{-1}$$

The material parameters are given as $c_{10} = c_1 = 0.916\,\text{MPa}$ and $c_{01} = c_2 = 0.0647\,\text{MPa}$. Plot a graph of normal stress versus stretch ratio in the direction (1).

Problem 4.2.2

The deformation gradient of a simple shear test is given as follows:

$$(\mathbf{F}) = \begin{pmatrix} 1 & \gamma & 0 \\ 0 & 1 & 0 \\ 0 & 0 & 1 \end{pmatrix}$$

γ is the shear strain and in terms of stretch ratio $\lambda = l/l_0$, it takes the following form:

$$\gamma = \lambda - 1/\lambda$$

furthermore

$$\lambda_1 = \lambda, \lambda_2 = 1/\lambda, \lambda_3 = 1$$

Plot the Cauchy stress tensor versus stretch ratio if the sample for the testing used is a rubber block modelled as an incompressible Mooney–Rivlin material with $c_{10} = c_1 = 0.220\,\text{MPa}$ and $c_{01} = c_2 = 0.017\,\text{MPa}$. Determine the Cauchy stress if the sample is stretched to 300 % of its initial length.

Solution

Problem 4.2.3

Experimental data of an isotropic incompressible rubberlike animal skin material are sometimes correlated by an Ogden model. The strain energy function of an Ogden material is defined as follows:

$$\psi(\lambda_1, \lambda_2, \lambda_3) = \sum_{i=1}^{n} \frac{\mu_i}{\alpha_i} (\lambda_1^{\alpha_i} + \lambda_2^{\alpha_i} + \lambda_3^{\alpha_i} - 3)$$

The parameters $\lambda_1, \lambda_2, \lambda_3$ are the eigenvalues of the right Cauchy–Green tensor $\mathbf{C}$ and μ_i, α_i are the material constants. Derive the constitutive equation for this material relating the second Piola-Kirchhoff stress tensor and right Cauchy–Green tensor $\mathbf{C}$.

Solution

Given is the spectral form of the Cauchy–Green tensor:

$$(\mathbf{C}) = \begin{pmatrix} \lambda_1 & 0 & 0 \\ 0 & \lambda_2 & 0 \\ 0 & 0 & \lambda_3 \end{pmatrix}$$

Therefore, we have

$$(\mathbf{C}^{\alpha_i}) = \begin{pmatrix} \lambda_1^{\alpha_i} & 0 & 0 \\ 0 & \lambda_2^{\alpha_i} & 0 \\ 0 & 0 & \lambda_3^{\alpha_i} \end{pmatrix}$$

Now we can rewrite $\lambda_1^{\alpha_i} + \lambda_2^{\alpha_i} + \lambda_3^{\alpha_i}$ as a trace of the tensor $\mathbf{C}^i$ or as a double scalar product, so that the above strain energy function can be written as

$$\psi(\lambda_1, \lambda_2, \lambda_3) = \sum_{i=1}^{n} \frac{\mu_i}{\alpha_i} (\lambda_1^{\alpha_i} + \lambda_2^{\alpha_i} + \lambda_3^{\alpha_i} - 3)$$

$$= \psi(\mathbf{C}) = \sum_{i=1}^{n} \frac{\mu_i}{\alpha_i} (\mathbf{I} \cdot\cdot \, \mathbf{C}^i - 3)$$

From Eq. (4.14), the second Piola-Kirchhoff tensor is given as

$$\tilde{\mathbf{T}} = 2 \frac{\partial \psi(\mathbf{C})}{\partial \mathbf{C}}$$

In order to obtain the partial derivative $\partial \psi(\mathbf{C})/\partial \mathbf{C}$, we apply the directional Gateaux derivative $\delta \psi(\mathbf{C}; \mathbf{K})$ in the direction of arbitrary tensor $\mathbf{K}$. The Gateaux derivative of $\delta \psi(\mathbf{C}; \mathbf{K})$ is define as follows:

$$\delta\psi(\mathbf{C}; \mathbf{K}) = (\frac{\partial \psi}{\partial \mathbf{C}})^T \cdot \cdot \mathbf{K} = \lim_{\xi \to 0} \frac{1}{\xi}[\psi(\mathbf{C} + \xi\mathbf{K}) - \psi(\mathbf{C})]$$

$$\delta\psi(\mathbf{C}; \mathbf{K}) = \lim_{\xi \to 0} \frac{1}{\xi}[\sum_{i=1}^{n} \frac{\mu_i}{\alpha_i} (\mathbf{I} \cdot \cdot (\mathbf{C} + {}_{\xi}\mathbf{K})^i - 3) - \sum_{i=1}^{n} \frac{\mu_i}{\alpha_i} (\mathbf{I} \cdot \cdot \mathbf{C}^i - 3)]$$

After expansion and subtraction, we obtain

$$\delta\psi(\mathbf{C}; \mathbf{K}) = \sum_{i=1}^{n} \frac{\mu_i}{\alpha_i} [\alpha_i \mathbf{I} \cdot \cdot \mathbf{C}^{i-1} \cdot \mathbf{K} = \sum_{i=1}^{n} \frac{\mu_i}{\alpha_i} [\alpha_i \mathbf{K} \cdot \cdot \mathbf{C}^{\alpha_i - 1}]$$

$$= \mathbf{K} \cdot \cdot \sum_{i=1}^{n} \frac{\mu_i}{\alpha_i} [\alpha_i \mathbf{C}^{\alpha_i - 1}]$$

$$\delta\psi(\mathbf{C}; \mathbf{K}) = (\frac{\partial \psi}{\partial \mathbf{C}})^T \cdot \cdot \mathbf{K} = \mathbf{K} \cdot \cdot \sum_{i=1}^{n} \frac{\mu_i}{\alpha_i} [\alpha_i \mathbf{C}^{\alpha_i - 1}]$$

which implies

$$(\frac{\partial \psi}{\partial \mathbf{C}})^T = \sum_{i=1}^{n} \frac{\mu_i}{\alpha_i} [\alpha_i \mathbf{C}^{\alpha_i - 1}]$$

Now the second Piola-Kirchhoff stress tensor is

$$\tilde{\mathbf{T}} = 2\sum_{i=1}^{n} \mu_i (\mathbf{C}^T)^{\alpha_i - 1} = 2\sum_{i=1}^{n} \mu_i \mathbf{C}^{\alpha_i - 1}$$

or

$$(1/2)\tilde{\mathbf{T}} = \mu_i \mathbf{I} + \mu_2 \mathbf{C} + \mu_3 \mathbf{C}^2 + \mu_4 \mathbf{C}^3 ...$$

which can be reduced to a linear combination of $\mathbf{I}, \mathbf{C}, \mathbf{C}^2$ or $\mathbf{I}, \mathbf{C}, \mathbf{C}^{-1}$ with appropriate coefficient functions of $\mathbf{C}$ by the Cayley–Hamilton theorem (Fig. 6.6).

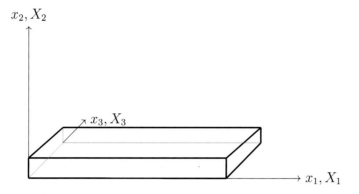

Fig. 6.6 Uniaxial tension test

Problem 4.2.4

A modified version of St. Venant-Kirchhoff constitutive behaviour is defined by the following strain-energy density function ψ as a scalar function of the Green–Lagrange strain tensor **E** and the Jacobian of the deformation gradient J

$$\Psi(\mathbf{E}, J) = \mu\, tr(\mathbf{E}^2) + \frac{\kappa}{2}(\ln J)^2$$

μ and κ are material constants.

1. Obtain an expression for the second Piola-Kirchhoff stress tensor $\hat{\mathbf{T}}$ as a function of the right Cauchy–Green tensor **C**.
2. Obtain an expression for first Piola-Kirchhoff stress tensor **T** as a function of the left Cauchy–Green tensor **B**.
3. Determine the material elasticity tensor.

Solution

Problem 4.2.5

Beginning from the generalized Hooke's law

$$\mathbf{S}(\mathbf{x}, t) = \underline{\mathbf{E}} : \hat{\mathbf{E}}(\mathbf{x}, t) \longleftrightarrow \sigma_{ij}(x_i, t) = E_{ijkl}\epsilon_{kl}(x_i, t) \tag{6.3}$$

derive the associated strain energy function ψ.

Solution

Problem 4.2.6

A rubber block is subjected to a simple shear test with the deformation gradient **F**

$$\mathbf{F} = \begin{pmatrix} 1 & \gamma & 0 \\ 0 & 1 & 0 \\ 0 & 0 & 1 \end{pmatrix}$$

γ is the shear strain. The material property of the rubber is modelled by a two-parameter Mooney–Rivlin hyperelastic material model with the strain energy density given as follows:

$$W = c_{10}(\hat{I}_1 3) + c_{01}(\hat{I}_2 - 3) + (1/D_1)(J - 1)^2$$

J is the Jacobian parameter. Taking the following data into consideration, calculate the resulting second Piola-Kirchhoff stress tensor:

$$c_{10} = -0.55\,\text{MPa},\ c_{01} = 0.7\,\text{MPa},\ D_1 = 0.001(\text{MPa})^{-1}$$

Solution

Problem 4.2.7

A thin sheet of an incompressible hyperelastic material is subjected to a biaxial tension test with the state of plane stress, i.e. $\hat{t}_{23} = \hat{t}_{13} = \hat{t}_{33} = 0$; show that

$$\hat{t}_{11} = 2\left(\frac{\partial\psi}{\partial I_{\mathbf{C}}} + (\lambda_1^2 + \lambda_2^2 + \frac{1}{\lambda_1^2\lambda_2^2})\frac{\partial\psi}{\partial II_{\mathbf{C}}}\right) - \lambda_1^2\frac{\partial\psi}{\partial II_{\mathbf{C}}}$$

$$\hat{t}_{22} = 2\left(\frac{\partial\psi}{\partial I_{\mathbf{C}}} + (\lambda_1^2 + \lambda_2^2 + \frac{1}{\lambda_1^2\lambda_2^2})\frac{\partial\psi}{\partial II_{\mathbf{C}}}\right) - \lambda_2^2\frac{\partial\psi}{\partial II_{\mathbf{C}}}$$

where $\hat{t}_{11} = \hat{t}_1$ and $\hat{t}_{22} = \hat{t}_2$ are the elements of second Piola-Kirchhoff stress in the principal axes and the $I_{\mathbf{C}}$ and $II_{\mathbf{C}}$ are the invariants of the right Cauchy–Green deformation tensor **C**. The right and left stretch tensor of a biaxial tension is given as follows:

$$\mathbf{F} = \mathbf{U} = \mathbf{V} = \begin{pmatrix} \lambda_1 & 0 & 0 \\ 0 & \lambda_2 & 0 \\ 0 & 0 & \frac{1}{\lambda_1\lambda_2} \end{pmatrix}$$

Solution

From the given right stretch tensor **U**, we calculate the right Cauchy–Green tensor
$\mathbf{C} = \mathbf{U}^2$

$$\mathbf{C} = \mathbf{U}^2 = \begin{pmatrix} \lambda_1^2 & 0 & 0 \\ 0 & \lambda_2^2 & 0 \\ 0 & 0 & \frac{1}{\lambda_1^2 \lambda_2^2} \end{pmatrix}$$

The two invariants $I_\mathbf{C}$ and $II_\mathbf{C}$ can obtained easily

$$I_\mathbf{C} = \lambda_1^2 + \lambda_2^2 + \frac{1}{\lambda_1^2 \lambda_2^2}$$

and

$$II_\mathbf{C} = \lambda_1^2 \lambda_2^2 + \frac{1}{\lambda_1^2} + \frac{1}{\lambda_2^2}$$

Now the constitutive equation in terms of the second Piola-Kirchhoff stress tensor and the right Cauchy–Green deformation tensor is given as follows:

$$\hat{\mathbf{T}} = 2\frac{\partial \psi}{\partial \mathbf{C}} = 2\frac{\partial \psi}{\partial I_\mathbf{C}} \cdot \frac{\partial I_\mathbf{C}}{\partial \mathbf{C}} + 2\frac{\partial \psi}{\partial II_\mathbf{C}} \cdot \frac{\partial II_\mathbf{C}}{\partial \mathbf{C}}$$

$$= 2\frac{\partial \psi}{\partial I_\mathbf{C}} \cdot \mathbf{I} + 2\frac{\partial \psi}{\partial II_\mathbf{C}} \cdot (I_\mathbf{C}\mathbf{I} - \mathbf{C})$$

$$= 2\left(\frac{\partial \psi}{\partial I_\mathbf{C}} + I_\mathbf{C}\frac{\partial \psi}{\partial II_\mathbf{C}} \right)\mathbf{I} - \frac{\partial \psi}{\partial II_\mathbf{C}}\mathbf{C}$$

After the substitution of **C**, **I** and $I_\mathbf{C}$ in the above equation, we obtain

$$\hat{t}_{11} = 2\left(\frac{\partial \psi}{\partial I_\mathbf{C}} + (\lambda_1^2 + \lambda_2^2 + \frac{1}{\lambda_1^2 \lambda_2^2})\frac{\partial \psi}{\partial II_\mathbf{C}} \right) - \lambda_1^2 \frac{\partial \psi}{\partial II_\mathbf{C}}$$

$$\hat{t}_{22} = 2\left(\frac{\partial \psi}{\partial I_\mathbf{C}} + (\lambda_1^2 + \lambda_2^2 + \frac{1}{\lambda_1^2 \lambda_2^2})\frac{\partial \psi}{\partial II_\mathbf{C}} \right) - \lambda_2^2 \frac{\partial \psi}{\partial II_\mathbf{C}}$$

6.16   Section 5.1

Problem 5.1.1

Given is the Cauchy stress tensor $\mathbf{S}$ as follows:

$$(\mathbf{S}) = \begin{pmatrix} x_1 + x_2 & f(x_1, x_2) & 0 \\ f(x_1, x_2) & x_1 - 2x_2 & 0 \\ 0 & 0 & x_2 \end{pmatrix}$$

1. Determine the function $f(x_1, x_2)$ assuming that the body force is negligibly small and the origin is stress-free.
2. Determine the stress vector at any point on the surface $x_1 = constant$ and $x_2 = constant$.

Solution

By substitution in the first Cauchy equation of motion, i.e. the principle of conversation of linear momentum, $\nabla \cdot \mathbf{S} = \mathbf{0}$ in the absence of body forces, we have

$$\nabla \cdot \mathbf{S} = \nabla \cdot \begin{pmatrix} x_1 + x_2 & f(x_1, x_2) & 0 \\ f(x_1, x_2) & x_1 - 2x_2 & 0 \\ 0 & 0 & x_2 \end{pmatrix} = \mathbf{0}$$

After carrying out the divergence operation, we obtain

$$\begin{pmatrix} \partial_1(x_1 + x_2) + \partial_2 f(x_1, x_2) \\ \partial_1 f(x_1, x_2) + \partial_2(x_1 - 2x_2) \\ 0 \end{pmatrix} = \begin{pmatrix} 0 \\ 0 \\ 0 \end{pmatrix}$$

with $\partial_1(.) = \partial(.)/\partial x_1$ and $\partial_2(.) = \partial(.)/\partial x_2$. Now we have two equations for determining the function $f(x_1, x_2)$:

$$1 + \partial f(x_1, x_2)/\partial x_2 = 0$$

and

$$\partial f(x_1, x_2)/\partial x_1 - 2 = 0$$

Integrating both equations over x_1 and x_2, we have

$$f(x_1, x_2) = -2x_1 + x_2 + C$$

Since the body is stress-free at the origin, then $C = 0$. Therefore, the function f is

$$f(x_1, x_2) = -2x_1 + x_2$$

Problem 5.1.2

If $\mathbf{T}$ is any symmetrical tensor field, i.e. $\mathbf{T} = \mathbf{T}(x_i)$, $i = 1, 2, 3$, is then $\mathbf{S}$, with

$$\mathbf{S} = \nabla \times (\nabla \times \mathbf{T})$$

statically allowable to represent the Cauchy stress tensor.

Solution

In the case of static and in the absence of body forces, the equation of motion, i.e. Cauchy I, takes the following form:

$$\nabla \cdot \mathbf{S} = \mathbf{0}$$

with $\mathbf{S}$ the Cauchy stress tensor. Then any symmetrical tensor $\mathbf{T} = \mathbf{T}(x_i)$ with

$$\mathbf{S} = \nabla \times (\nabla \times \mathbf{T})$$

can represent the Cauchy stress tensor if $\nabla \cdot (\nabla \times (\nabla \times \mathbf{T}))$ is identically zero vector. Firstly, we use the identity

$$\mathbf{a} \times (\mathbf{b} \times \mathbf{c}) = (\mathbf{a} \cdot \mathbf{c})\mathbf{b} - (\mathbf{a} \cdot \mathbf{b})\mathbf{c}$$

to simplify the above expression for $\mathbf{S}$ and obtain

$$\mathbf{S} = \nabla(\nabla \cdot \mathbf{T}) - (\nabla \cdot \nabla)\mathbf{T}$$

The divergence of $\mathbf{S}$ is then

$$\nabla \cdot \mathbf{S} = \nabla \cdot (\nabla(\nabla \cdot \mathbf{T}) - (\nabla \cdot \nabla)\mathbf{T})$$

After expansion, it becomes

$$\nabla \cdot \nabla(\nabla \cdot \mathbf{T}) - (\nabla \cdot \nabla)(\nabla \cdot \mathbf{T})$$

which is clearly a zero vector.

Problem 5.1.3

Show that the Cauchy stress tensor possesses only real eigenvalues

Solution

Assume that λ is the eigenvalues of the symmetrical Cauchy stress tensor $\mathbf{S} = \mathbf{S}^T$ and $\mathbf{x}$ is the associated eigenvector, therefore

$$\mathbf{S} \cdot \mathbf{x} = \lambda \mathbf{x}, \quad \mathbf{x} \neq \mathbf{0}$$

Now with $\mathbf{x}^*$ and $\mathbf{S}^*$, the conjugate of $\mathbf{x}$ and $\mathbf{S}$ respectively, we write

$$\lambda \mathbf{x}^{*T} \cdot \mathbf{x} = \mathbf{x}^{*T} \cdot (\lambda \mathbf{x}), \quad \mathbf{x}^{*T} \cdot \mathbf{x} \neq \mathbf{0}$$

Rearrange further
$$= \mathbf{x}^{*T} \cdot (\mathbf{S} \cdot \mathbf{x})$$

$$= (\mathbf{S}^T \cdot \mathbf{x}^*)^T \cdot \mathbf{x} = (\mathbf{S} \cdot \mathbf{x}^*)^T \cdot \mathbf{x} = (\mathbf{S}^* \cdot \mathbf{x}^*)^T \cdot \mathbf{x}$$

$$= (\lambda^* \mathbf{x}^*)^T \cdot \mathbf{x}$$

We have then
$$\lambda \mathbf{x}^{*T} \cdot \mathbf{x} = (\lambda^* \mathbf{x}^*)^T \cdot \mathbf{x} = \lambda^* (\mathbf{x}^{*T} \cdot \mathbf{x})$$

which implies
$$\lambda = \lambda^*$$

Any number that equals its conjugate is a real number. Therefore, a symmetrical Cauchy stress tensor possesses real eigenvalues.

Problem 5.1.4

A motion is referred to as potential motion if its velocity $\mathbf{v}$ can be expressed as

$$\mathbf{v} = \nabla \phi$$

with ϕ a scalar function of position.
    Derive the continuity equation for such motion of a compressible and an incompressible medium in terms of the potential ϕ.

Solution

We begin the Lagrangian formulation of the continuity equation, which is

$$\frac{\partial \rho}{\partial t} + \nabla \cdot (\rho \mathbf{v}) = 0$$

Now for the potential motion, $\mathbf{v} = \nabla \phi$, we have then for incompressible material $\frac{\partial \rho}{\partial t} = 0$, which yields

$$\nabla \cdot (\rho \nabla \phi) = 0$$

or for a homogeneous material with position-independent density

$$\nabla \cdot (\nabla \phi) = 0$$

$$\Delta \phi = 0, \quad \Delta = \nabla \cdot \nabla$$

The above equation is well known as the Laplace equation of ϕ. In Lagrangian or material coordinates, the equation takes the following form:

$$\frac{\partial^2 \phi}{\partial^2 X} + \frac{\partial^2 \phi}{\partial^2 Y} + \frac{\partial^2 \phi}{\partial^2 Z} = 0$$

or for very small motion

$$\frac{\partial^2 \phi}{\partial^2 x} + \frac{\partial^2 \phi}{\partial^2 y} + \frac{\partial^2 \phi}{\partial^2 z} = 0$$

In the case of compressible material, the continuity equation for potential motion is

$$\frac{\partial \rho}{\partial t} + \rho \left(\frac{\partial^2 \phi}{\partial^2 x} + \frac{\partial^2 \phi}{\partial^2 y} + \frac{\partial^2 \phi}{\partial^2 z} \right) = 0$$

Problem 5.1.5

Given is the velocity field of a two-dimensional flow as follows:

$$\mathbf{v}(x, y) = (ax + b)\mathbf{e_x} + (-ay + c)\mathbf{e_y}$$

Find the potential function ϕ of the flow and verify that it obeys the Laplace equation.

Solution

With the definition of the potential function $\phi(x, y)$, the component of the velocity field can be written as

$$v_x = ax + b = \partial\phi/\partial x$$

$$v_y = -ay + c = \partial\phi/\partial y$$

The integration of both equations above over x and y respectively yields

$$\phi = ax^2 + bx + f(y) + C \quad (*)$$

where C is the integration constant. Again differentiation of the above equation with respect to y produces

$$\partial\phi/\partial y = df(y)/dy = v_y = -ay + c$$

or after integration over y,

$$f(y) = -ay^2 + cy + D$$

Substituting back into Eq. (*), we have the potential function ϕ

$$\phi(x, y) = ax^2 + bx - ay^2 + cy + D + C$$

or after simplification

$$\phi(x, y) = ax^2 - ay^2 + +bx + cy + E$$

It is obvious that the function above fulfils the Laplace equation.

Problem 5.1.6

Show that the planar state of stress can be described by the following equation:

$$\sigma_{11} = \psi_{,22} + \kappa, \ \sigma_{22} = \psi_{,11} + \kappa, \ \sigma_{12} = -\psi_{,12}$$

whereby $\psi(x_1, x_2)$ is the stress function (Airy's stress function) and κ is a potential field of the body force $\mathbf{f_b}$, with $\nabla\kappa = \mathbf{f_b}$.

Solution

Starting from the equilibrium equation, i.e. equation of motion for the static case,

$$\nabla \cdot \mathbf{S} + \mathbf{f_b} = \mathbf{0}$$

or componentwise,

$$\sigma_{11,1} + \sigma_{12,2} + f_{b1} = 0$$

$$\sigma_{21,1} + \sigma_{22,2} + f_{b2} = 0$$

We have after substituting the above equations the following:

$$\psi_{,221} + \kappa_{,1} - \psi_{,122} = 0$$

$$-\psi_{,121} + \psi_{,112} + \kappa_{,2} = 0$$

Since the gradient of the potential function κ is defined as a body force

$$\kappa_{,1} = f_{b1}, \quad \kappa_{,2} = f_{b2}$$

then the above equations with the Airy function satisfied the equilibrium equation automatically, hence they are suitable to represent the planar state of stress.

Appendix

I. Multiplication Between Vectors and Tensors

In tensor algebra besides the common scalar and vector (or exterior) products, there are two more multiplicative operations between these quantities. They are a dyadic product and a double scalar product.

(a) Vector Product: Epsilon Tensor

Let $\mathbf{a} = a_q \mathbf{e}_q$ and $\mathbf{b} = b_r \mathbf{e}_r$ be two Cartesian vectors. The vector product between $\mathbf{a}$ and $\mathbf{b}$ is defined as

$$\mathbf{a} \times \mathbf{b} = \epsilon_{pqr} a_q b_r \mathbf{e}_p$$

where ϵ_{pqr} is the component of a third-order tensor. This tensor is called epsilon or Levi–Civita permutation tensor:

$$\epsilon_{pqr} = \begin{cases} 1 & \text{cyclic} \\ -1 & \text{acyclic} \\ 0 & \text{otherwise} \end{cases}$$

The terms "cyclic" and "acyclic" refer to the even and odd permutations of indices pqr.

(b) Dyadic Product

Let $\mathbf{a} = a_i \mathbf{e}_i$ and $\mathbf{b} = b_j \mathbf{e}_j$ be two Cartesian vectors. Then the dyadic product between $\mathbf{a}$ and $\mathbf{b}$ is defined as

$$\mathbf{a} \otimes \mathbf{b} = a_i b_j \mathbf{e}_i \otimes \mathbf{e}_j$$

With Einstein's convention, it takes the form

N. A. N. Mohamed, *Introduction to Continuum Mechanics for Engineers*, https://doi.org/10.1007/978-981-99-0811-0

$$\mathbf{a} \otimes \mathbf{b} = a_1 b_1 \, \mathbf{e}_1 \otimes \mathbf{e}_1 + a_1 b_2 \, \mathbf{e}_1 \otimes \mathbf{e}_2 + a_1 b_3 \, \mathbf{e}_1 \otimes \mathbf{e}_3$$
$$+ \, a_2 b_1 \, \mathbf{e}_2 \otimes \mathbf{e}_1 + a_2 b_2 \mathbf{e}_2 \otimes \mathbf{e}_2 + a_2 b_3 \, \mathbf{e}_2 \otimes \mathbf{e}_3$$
$$+ \, a_3 b_1 \, \mathbf{e}_3 \otimes \mathbf{e}_1 + a_3 b_2 \, \mathbf{e}_3 \otimes \mathbf{e}_2 + a_3 b_3 \, \mathbf{e}_3 \otimes \mathbf{e}_3$$

Example

Let $\mathbf{a} = 2\mathbf{e}_1 + \mathbf{e}_2 - 5\mathbf{e}_3$ and $\mathbf{b} = \mathbf{e}_1 + \mathbf{e}_2 + 5\mathbf{e}_3$. Then the dyadic product between **a** and **b** is

$$\mathbf{a} \otimes \mathbf{b} = 2\,\mathbf{e}_1 \otimes \mathbf{e}_1 + 2\,\mathbf{e}_1 \otimes \mathbf{e}_2 + 10\,\mathbf{e}_1 \otimes \mathbf{e}_3$$
$$+ \quad \mathbf{e}_2 \otimes \mathbf{e}_1 + \mathbf{e}_2 \otimes \mathbf{e}_2 + 5\,\mathbf{e}_2 \otimes \mathbf{e}_3$$
$$- \, 5\,\mathbf{e}_3 \otimes \mathbf{e}_1 - 5\,\mathbf{e}_3 \otimes \mathbf{e}_2 - 25\,\mathbf{e}_3 \otimes \mathbf{e}_3$$

(c) Double Scalar Product

The double scalar product of two second-order tensors is a contraction of those tensors to a scalar quantity. Let $\mathbf{S} = S_{ij} \, \mathbf{e}_i \otimes \mathbf{e}_j$ and $\mathbf{T} = T_{kl} \, \mathbf{e}_k \otimes \mathbf{e}_l$ be any two Cartesian tensors of second order, then the double scalar product is defined as either

$$A) \quad \mathbf{S} \cdot \cdot \mathbf{T} = S_{ij} T_{ji}$$

or

$$B) \quad \mathbf{S} : \mathbf{T} = S_{ij} T_{ij}$$

The first one is the horizontal double scalar product, whereas the second one is the vertical double scalar product. As one can see

$$\mathbf{S} : \mathbf{T} = \mathbf{S} \cdot \cdot \mathbf{T}^T$$

II. Voigt Notation

One way of simplifying the common use of indicial tensor notation in continuum mechanics is to rearrange the tensor quantities as a matrix operation, thus avoiding the use of the lengthy Einstein convention of summation. For tensors of second order, this task is relatively easy, not however for the tensors of higher orders. For instance, for the material equation for linear elastic material

$$\sigma_{ij} = E_{ijkl} \, \epsilon_{kl}$$

the tensor σ_{ij} and ϵ_{kl} is written as column vectors 1×6 and the material tensor E_{ijkl} as a square matrix 6×6

$$
\begin{pmatrix} \sigma_1 \\ \sigma_2 \\ \sigma_3 \\ \sigma_4 \\ \sigma_5 \\ \sigma_6 \end{pmatrix} = \begin{pmatrix} E_{11} & E_{12} & E_{13} & E_{14} & E_{15} & E_{16} \\ & E_{22} & E_{23} & E_{24} & E_{25} & E_{26} \\ & & E_{33} & E_{34} & E_{35} & E_{36} \\ & & & E_{44} & E_{45} & E_{46} \\ & symm & & & E_{55} & E_{56} \\ & & & & & E_{66} \end{pmatrix} \begin{pmatrix} \epsilon_1 \\ \epsilon_2 \\ \epsilon_3 \\ \epsilon_4 \\ \epsilon_5 \\ \epsilon_6 \end{pmatrix}
$$

The matrix arrangement above is known as Voigt notation or engineering notation. The reduction of the number of elements in the material tensor from 81 to 36 is owing to the symmetry of both Cauchy stress and Lagrangian strain tensors. Further arrangements are

- combined index pairs of normal components $()_{11} \to ()_1$, $()_{22} \to ()_2$ and $()_{33} \to ()_3$, of tangential elements $()_{23} \to ()_4$, $()_{13} \to ()_5$ and $()_{12} \to ()_6$.
- defining the engineering shear strain as a sum of two symmetric components, that is, $2\epsilon_{23} = \epsilon_{23} + \epsilon_{32} \leftarrow \epsilon_4$, etc.

.

III. Gâteaux Derivative

Let $\mathcal{X}$ and $\mathcal{Y}$ be linear spaces over real number $\mathbf{R}$ and A a linear mapping from $\mathcal{X}$ to $\mathcal{Y}$. The mapping A is said to be Gâteaux differentiable at the element $\mathbf{T} \in \mathcal{X}$, if there exists a linear operator $\mathbf{L}$ with the following property:

$$
A(\mathbf{T} + \xi \mathbf{K}) - A(\mathbf{T}) = \xi L(\mathbf{K}) + \mathcal{O}(\xi), \xi \in \mathbf{R}, \ \xi \to \iota
$$

for all $\mathbf{K} \in \mathcal{X}$ from the neighbourhood of $\mathbf{T}$. The linear operator L is called the Gâteaux derivative or Gâteaux variation, and it can be understood as the directional differential of the mapping A at the element $\mathbf{T}$ in the direction of the element $\mathbf{K} \in \mathcal{X}$. In practical calculation, the above definition equation can be rewritten as follows:

$$
L(\mathbf{K}) = \delta A(\mathbf{T}; \mathbf{K}) = \lim_{\xi \to 0} \frac{1}{\xi} [A(\mathbf{T} + \xi \mathbf{K}) - A(\mathbf{T})]
$$

The directional derivative $\delta A(\mathbf{T}; \mathbf{K})$ can be split into the partial derivative of the mapping A with respect to $\mathbf{T}$ in the direction of $\mathbf{K} \in \mathcal{X}$ as follows:

$$
\delta A(\mathbf{T}; \mathbf{K}) = (\frac{\partial A}{\partial \mathbf{T}})^T \cdot \cdot \mathbf{K}
$$

Example

The partial derivative of the second invariant of $\mathbf{C}$ with respect to $\mathbf{C}$. The second invariant of $\mathbf{C}$, $II_{\mathbf{C}}$ is given by the following equation:

$$IIc = \frac{1}{2}((tr\ C)^2 - tr\ C^2) = \frac{1}{2}((\mathbf{I} \cdot \cdot \mathbf{C})(\mathbf{I} \cdot \cdot \mathbf{C}) - (\mathbf{C} \cdot \cdot \mathbf{C}))$$

The Gâteaux derivative of IIc is according to the definition

$$\delta IIc(\mathbf{C}; \mathbf{K}) = \lim_{\xi \to 0} \frac{1}{\xi}[IIc(\mathbf{C} + \xi \mathbf{K}) - IIc(\mathbf{C})]$$

$$\frac{1}{\xi}[IIc(\mathbf{C} + \xi \mathbf{K}) - IIc(\mathbf{C})] =$$

$$\frac{1}{2\xi}(\mathbf{I} \cdot \cdot (\mathbf{C} + \xi \mathbf{K}))(\mathbf{I} \cdot \cdot (\mathbf{C} + \xi \mathbf{K})) - \frac{1}{2\xi}\mathbf{I} \cdot \cdot (\mathbf{C} + \xi \mathbf{K}) \cdot (\mathbf{C} + \xi \mathbf{K}) - \frac{1}{2\xi}((\mathbf{I} \cdot \cdot \mathbf{C})(\mathbf{I} \cdot \cdot \mathbf{C}) - (\mathbf{C} \cdot \cdot \mathbf{C}))$$

$$= \frac{1}{2\xi}[\xi(\mathbf{I} \cdot \cdot \mathbf{K})(\mathbf{I} \cdot \cdot \mathbf{C}) + \xi(\mathbf{I} \cdot \cdot \mathbf{C})(\mathbf{I} \cdot \cdot \mathbf{K})] - \frac{1}{2\xi}[\xi \mathbf{I} \cdot \cdot \mathbf{C} \cdot \mathbf{K} + \xi \mathbf{I} \cdot \cdot \mathbf{K} \cdot \mathbf{C}]$$

$$- \frac{1}{2\xi}((\mathbf{I} \cdot \cdot \xi \mathbf{K})(\mathbf{I} \cdot \cdot \xi \mathbf{K}) - (\xi \mathbf{K} \cdot \cdot \xi \mathbf{K}))$$

As $\xi \to 0$, the above expression for the Gâteux derivative simplifies to

$$= \frac{1}{2}[(\mathbf{I} \cdot \cdot \mathbf{K})(\mathbf{I} \cdot \cdot \mathbf{C}) + (\mathbf{I} \cdot \cdot \mathbf{C})(\mathbf{I} \cdot \cdot \mathbf{K})] - \frac{1}{2}[\mathbf{I} \cdot \cdot \mathbf{C} \cdot \mathbf{K} + \mathbf{I} \cdot \cdot \mathbf{K} \cdot \mathbf{C}]$$

Since $(\mathbf{I} \cdot \cdot \mathbf{K})(\mathbf{I} \cdot \cdot \mathbf{C}) = (\mathbf{I} \cdot \cdot \mathbf{C})(\mathbf{I} \cdot \cdot \mathbf{K})$ and $\mathbf{I} \cdot \cdot \mathbf{C} \cdot \mathbf{K} = \mathbf{I} \cdot \cdot \mathbf{K} \cdot \mathbf{C} = \mathbf{K} \cdot \cdot \mathbf{C} = \mathbf{C} \cdot \cdot \mathbf{K}$
we have then

$$\delta IIc(\mathbf{C}; \mathbf{K}) = \lim_{\xi \to 0} \frac{1}{\xi}[IIc(\mathbf{C} + \xi \mathbf{K}) - IIc(\mathbf{C})]$$

$$= (\mathbf{I} \cdot \cdot \mathbf{C})(\mathbf{I} \cdot \cdot \mathbf{K}) - \mathbf{C} \cdot \cdot \mathbf{K}$$

Now with $\mathbf{I} \cdot \cdot \mathbf{C} = IIc$ the second invariant of the tensor $\mathbf{C}$ the above equation reduces to

$$= IIc \mathbf{I} \cdot \cdot \mathbf{K} - \mathbf{C} \cdot \cdot \mathbf{K}$$

$$= (IIc \mathbf{I} - \mathbf{C}) \cdot \cdot \mathbf{K}$$

or per definition

$$= (IIc \mathbf{I} - \mathbf{C}) \cdot \cdot \mathbf{K} = (\frac{\partial IIc}{\partial \mathbf{T}})^T \cdot \cdot \mathbf{K}$$

Therefore, for all $\mathbf{K} \in \mathcal{X}$ we have the partial derivative of the second invariant of $\mathbf{C}$ with respect to the tensor $\mathbf{C}$

$$\left(\frac{\partial II_C}{\partial \mathbf{T}}\right)^T = (II_C\mathbf{I} - \mathbf{C})$$

IV. Isotropic Tensor Function

Let $\mathbf{A} = \mathbf{F}(\mathbf{B})$ be any tensor-valued of the tensor function of a symmetrical tensor variable $\mathbf{B}$, then according to the Cayley–Hamilton theorem $\mathbf{A}$ can be expressed as a second-degree polynomial of argument tensor, $\mathbf{B}$, i.e.

$$\mathbf{A} = \phi_0\mathbf{I} + \phi_1\mathbf{B} + \phi_2\mathbf{B} \cdot \mathbf{B}$$

The scalar parameters ϕ_0, ϕ_1 and ϕ_2 are the functions of the irreducible invariants of the tensor $\mathbf{B}$. These functions are also called the integrity basis of the polynomial representation. The function $\mathbf{F}$ is then called isotropic, if the following condition of form-invariant is obeyed:

$$\mathbf{Q} \cdot \mathbf{F}(\mathbf{B}) \cdot \mathbf{Q}^T = \mathbf{F}(\mathbf{Q} \cdot \mathbf{B} \cdot \mathbf{Q}^T)$$

The above equation describes the orthogonal transformation of the function $\mathbf{F}$ with an orthogonal tensor $\mathbf{Q}$. If this transformation is obeyed, then the tensor function $\mathbf{F}$ is called an isotropic tensor function.

V. Falk's Scheme of Matrix Multiplication

Falk's scheme is a very useful and simple method of performing matrix multiplication between two or more matrices:

$$\mathbf{A} = \begin{pmatrix} a_{11} & a_{12} & a_{13} \\ a_{21} & a_{22} & a_{23} \\ a_{31} & a_{32} & a_{33} \end{pmatrix}$$

$$\mathbf{B} = \begin{pmatrix} b_{11} & b_{12} & b_{13} \\ b_{21} & b_{22} & b_{23} \\ b_{31} & b_{32} & b_{33} \end{pmatrix}$$

$$\mathbf{C} = \mathbf{A} \cdot \mathbf{B} = \begin{pmatrix} & & & b_{11} & b_{12} & b_{13} \\ & & & b_{21} & b_{22} & b_{23} \\ & & & b_{31} & b_{32} & b_{33} \\ a_{11} & a_{12} & a_{13} & c_{11} & c_{12} & c_{13} \\ a_{21} & a_{22} & a_{23} & c_{21} & c_{22} & c_{23} \\ a_{31} & a_{32} & a_{33} & c_{31} & c_{32} & c_{33} \end{pmatrix}$$

$$c_{ij} = \sum_k a_{ik} b_{kj}$$

for example

$$c_{11} = a_{11}b_{11} + a_{12}b_{21} + a_{13}b_{31}$$

$$.....$$

$$c_{23} = a_{21}b_{13} + a_{22}b_{23} + a_{23}b_{33}$$

$$.....$$

$$c_{33} = a_{31}b_{13} + a_{32}b_{23} + a_{33}b_{33}$$

$$\mathbf{D} = \mathbf{B} \cdot \mathbf{A} = \begin{pmatrix} & & & a_{11}\ a_{12}\ a_{13} \\ & & & a_{21}\ a_{22}\ a_{23} \\ & & & a_{31}\ a_{32}\ a_{33} \\ b_{11}\ b_{12}\ b_{13} & d_{11}\ d_{12}\ d_{13} \\ b_{21}\ b_{22}\ b_{23} & d_{21}\ d_{22}\ d_{23} \\ b_{31}\ b_{32}\ b_{33} & d_{31}\ d_{32}\ d_{33} \end{pmatrix}$$

$$d_{ij} = \sum_k b_{ik} a_{kj}$$

for example

$$d_{11} = b_{11}a_{11} + b_{12}a_{21} + b_{13}a_{31}$$

$$.....$$

$$d_{23} = b_{21}a_{13} + b_{22}a_{23} + b_{23}a_{33}$$

$$.....$$

$$d_{33} = b_{31}a_{13} + b_{32}a_{23} + b_{33}a_{33}$$

$$\mathbf{E} = \mathbf{B} \cdot \mathbf{A} \cdot \mathbf{B} = \mathbf{B} \cdot \mathbf{C} = \begin{pmatrix} & & & c_{11}\ c_{12}\ c_{13} \\ & & & c_{21}\ c_{22}\ c_{23} \\ & & & c_{31}\ c_{32}\ c_{33} \\ b_{11}\ b_{12}\ b_{13} & e_{11}\ e_{12}\ e_{13} \\ b_{21}\ b_{22}\ b_{23} & e_{21}\ e_{22}\ e_{23} \\ b_{31}\ b_{32}\ b_{33} & e_{31}\ e_{32}\ e_{33} \end{pmatrix}$$

$$
\mathbf{E} = \mathbf{B} \cdot \mathbf{A} \cdot \mathbf{B} = \mathbf{B} \cdot \mathbf{C} =
\begin{pmatrix}
 & & & & a_{11}\ a_{12}\ a_{13} \\
 & & & & a_{21}\ a_{22}\ a_{23} \\
 & & & & a_{31}\ a_{32}\ a_{33} \\
b_{11}\ b_{12}\ b_{13}\ & c_{11}\ c_{12}\ c_{13} \\
b_{21}\ b_{22}\ b_{23}\ & c_{21}\ c_{22}\ c_{23} \\
b_{31}\ b_{32}\ b_{33}\ & c_{31}\ c_{32}\ c_{33} \\
b_{11}\ b_{12}\ b_{13}\ & e_{11}\ e_{12}\ e_{13} \\
b_{21}\ b_{22}\ b_{23}\ & e_{21}\ e_{22}\ e_{23} \\
b_{31}\ b_{32}\ b_{33}\ & e_{31}\ e_{32}\ e_{33}
\end{pmatrix}
$$

or

$$
\mathbf{E} = \mathbf{B} \cdot \mathbf{A} \cdot \mathbf{B} = \mathbf{D} \cdot \mathbf{B} =
\begin{pmatrix}
 & & a_{11}\ a_{12}\ a_{13}\ b_{11}\ b_{12}\ b_{13} \\
 & & a_{21}\ a_{22}\ a_{23}\ b_{21}\ b_{22}\ b_{23} \\
 & & a_{31}\ a_{32}\ a_{33}\ b_{31}\ b_{32}\ b_{33} \\
b_{11}\ b_{12}\ b_{13}\ & d_{11}\ d_{12}\ d_{13}\ e_{11}\ e_{12}\ e_{13} \\
b_{21}\ b_{22}\ b_{23}\ & d_{21}\ d_{22}\ d_{23}\ e_{11}\ e_{12}\ e_{13} \\
b_{31}\ b_{32}\ b_{33}\ & d_{31}\ d_{32}\ d_{33}\ e_{11}\ e_{12}\ e_{13}
\end{pmatrix}
$$

References

N. Abdullah, O. Kastner, I. Mueller, A. Musolff, H. Xu, G. Zak, Observations on CuAlNi single crystals. Int. J. Non-Linear Mech. **37**(2002), 1263–1274 (2002)

S. Bechtel, R. Lowe, *Fundamental of Continuum Mechanics* (Elsevier Inc.; Academic, London, 2014)

E. Becker u. W. Buerger., *Kontinuumsmechanik* (Tuebner Verlag. Stuttgart, Germany, 1975)

J. Betten, *Kontinuumsmechanik- Elasto Plasto und Kriechmechanik* (Springer, Berlin, 1993)

P. Chadwick, *Thermomechanics of Rubberlike Materials*. Philosophical Transactions of the Royal Society of London, Series A, vol. 276 (1974)

P. Chadwick, *Continuum Mechanics: Concise Theory and Problems* (Dover, Illinois, 1999)

A.C. Eringen, *Continuum Physics*, vol. I (Academic, New York, 1971), pp 1–155

S. Hawking, *A Brief History of Time* (Bantam Dell Publishing Group (Bantam Books), London, 1988)

G.A. Holzapfel, R. Ogden, Constitutive modelling of passive myocardium: a structurally based framework for material characterization. Philos. Trans. R. Soc. **367**, 3445–3475 (2009)

W.M. Lai, D. Rubin, E. Krempl, *Introduction to Continuum Mechanics* (Butterworth - Heinemann, UK, 2009)

N.A.N. Mohamed, *Konjugierte Spannungstensoren und die Anwendung in der Kontinuumsmechanik* (VDI- Verlag. Duesseldorff, Germany, 1991)

N.A.N. Mohamed, *Analisis Tensor Kejuruteraan* (Penerbit UKM. Bangi, Malaysia, 2001)

S. Murakami, N. Ohno, *Creep in Structure* (Springer, Berlin, 1981), pp. 422–444

S. Murakami, *Continuum Damage Mechanics* (Springer, Vienna, 1987)

R.W. Ogden, Large deformation isotropic elasticity - on the correlation of theory and experiment for incompressible rubberlike solids **326**(1567) (1972). https://doi.org/10.1098/rspa.1972.0026

A.A. Shabana, *Computational Continuum Mechanics* (Cambridge University Press, London, 2012)

O.A. Shergold, N.A. Fleck, D. Radford, The uniaxial stress versus strain response of pig skin and silicone rubber at low and high strain rates. Int. J. Impact Eng. **32**, 1384–1402 (2006)

Sobolew, "Elemente der Funktionalanalysis" , H. Deutsch, Frankfurt a.M. (1979) (in German)

P. Thamburaja, N. Nikabdullah, A macroscopic constitutive model for shape-memory alloys: theory and finite-element simulations. Comput. Methods Appl. Mech. Eng. **198**(2009), 1074–1086 (2009)

© The Editor(s) (if applicable) and The Author(s), under exclusive license to Springer   185
Nature Singapore Pte Ltd. 2023
N. A. N. Mohamed, *Introduction to Continuum Mechanics for Engineers*,
https://doi.org/10.1007/978-981-99-0811-0

C. Truesdell, W. Noll, *Non-linear Field. Theories, of Mechanics* (Springer, Berlin, 1992)

C. Truesdell, *A First Course in Rational Continuum Mechanics* (Academic, New York, 1977)

R. Xiao, B. Hou, Q.P. Sun, H. Zhao, Y.L. Li, A numerical investigation of the nucleation and the propagation of NiTi martensitic transformation front under impact loading. Int. J. Impact Eng. **152** 103841 (2021)

Index

© The Editor(s) (if applicable) and The Author(s), under exclusive license to Springer
Nature Singapore Pte Ltd. 2023
N. A. N. Mohamed, *Introduction to Continuum Mechanics for Engineers*,
https://doi.org/10.1007/978-981-99-0811-0

Printed in the United States
by Baker & Taylor Publisher Services